Induction Motors

Edited by **Maurice Willis**

LANRYE
INTERNATIONAL

New Jersey

Published by Clanrye International,
55 Van Reypen Street,
Jersey City, NJ 07306, USA
www.clanryeinternational.com

Induction Motors
Edited by Maurice Willis

International Standard Book Number: 978-1-63240-304-9 (Hardback)

Printed in the United States of America.

Contents

Preface

Every book is a source of knowledge and this one is no exception. The idea that led to the conceptualization of this book was the fact that the world is advancing rapidly; which makes it crucial to document the progress in every field. I am aware that a lot of data is already available, yet, there is a lot more to learn. Hence, I accepted the responsibility of editing this book and contributing my knowledge to the community.

Prompted by the necessity of power-efficient progresses, procedure optimization, soft-start capacity and more ecological profits, there is a growing need to use induction motors for multiple functions at steady adaptable paces. These motors are capable of providing high manufacturing outcome with energy efficiency in varied industrial uses and are the core of advanced automation. This book disseminates knowledge about growing concerns in this area, like modeling. The collaboration of contributions by renowned scientists make this book not just a mechanical account of information but an international exploration on the technology of induction motors for readers. It provides them with a mixture of theory, implementation problems and practical examples.

While editing this book, I had multiple visions for it. Then I finally narrowed down to make every chapter a sole standing text explaining a particular topic, so that they can be used independently. However, the umbrella subject sinews them into a common theme. This makes the book a unique platform of knowledge.

I would like to give the major credit of this book to the experts from every corner of the world, who took the time to share their expertise with us. Also, I owe the completion of this book to the never-ending support of my family, who supported me throughout the project.

Editor

Control and Diagnosis

Tuning PI Regulators for Three-Phase Induction Motor Space Vector Modulation Direct Torque Control Using Complex Transfer Function Concept

Alfeu J. Sguarezi Filho, José L. Azcue P. and Ernesto Ruppert

Additional information is available at the end of the chapter

1. Introduction

The dynamics of induction motor (IM) is traditionally represented by differential equations. The space-vector concept [13] is used in the mathematical representation of IM state variables such as voltage, current, and flux.

The concept of complex transfer function derives from the application of the Laplace transform to differential equations in which the complex coefficients are in accordance with the spiral vector theory which has been presented by [24]. The complex transfer function concept is applied to the three-phase induction motor mathematical model and the induction motor root locus was presented in [10]. Other procedures for modeling and simulating the three-phase induction motor dynamics using the complex transfer function concept are also presented in [4].

The induction machine high performance dynamics is achieved by the field orientation control (FOC) [1, 17]. The three-phase induction motor field orientation control using the complex transfer function concept to tune the PI controller by using the frequency-response function of the closed-loop complex transfer function of the controlled induction machine was presented in [2]. This strategy has satisfactory current response although stator currents had presented cross-coupling during the induction machine transients. An interesting solution was presented in [11] in which it was designed a stator-current controller using complex form. From this, the current controller structure employing single-complex zeros is synthesized with satisfactory high dynamic performance although low-speed tests had not been shown in mentioned strategies.

An alternative for induction motor drive is the direct torque control (DTC), which consists of the direct control of the stator flux magnitude λ_1 and the electromagnetic torque T_e. DTC controllers generate a stator voltage vector that allows quick torque response with the smallest variation of the stator flux. The principles of the DTC using hysteresis controllers and variable switching frequency have been presented by [22] and [6]. It has disadvantages such as low speed operation [19].

The PI-PID controllers are widely used in control process in industry [18]. The PI controller was applied to the IM direct torque control has been presented by [23]. Some investigations to tune the PI gains of speed controller have been presented using genetic-fuzzy [20] and neural networks [21]. These strategies have satisfactory torque and flux response although a method to tune the PI controllers for stator flux and electromagnetic torque loop and low-speed tests had not been shown.

To overcome low speed operation shortcomings, various approaches for DTC applying flux vector acceleration method [9, 14] and deadbeat controller [5, 12, 15] have been reported. These strategies aim the induction motor control at low speed. In this case, the complex transfer function was not used to tune PI controllers for such strategy when the induction motor operates at any speed.

The aim of this book chapter is to provide the designing and tuning method for PI regulators, based on the three-phase induction motor mathematical model complex transfer function to be used in induction motor direct torque control when the machine operates at low speed which is a problem so far. This methods is in accordance with the present state of the art. The PI controller was designed and tuned by frequency-response function of the closed loop system. The controller also presents a minor complexity to induction motor direct torque control implementation. Experimental results are carried out to validate the controller design.

2. The complex model of the induction motor

The three-phase induction motor mathematical model in synchronous reference frame (dq) is given by [16]

$$\vec{v}_{1dq} = R_1\vec{i}_{1dq} + \frac{d\vec{\lambda}_{1dq}}{dt} + j\omega_1\vec{\lambda}_{1dq} \tag{1}$$

$$0 = R_2\vec{i}_{2dq} + \frac{d\vec{\lambda}_{2dq}}{dt} + j\left(\omega_1 - P\omega_{mec}\right)\vec{\lambda}_{2dq} \tag{2}$$

the relationship between fluxes and currents

$$\vec{\lambda}_{1dq} = L_1\vec{i}_{1dq} + L_M\vec{i}_{2dq} \tag{3}$$

$$\vec{\lambda}_{2dq} = L_M\vec{i}_{1dq} + L_2\vec{i}_{2dq} \tag{4}$$

The electromagnetic torque is expressed in terms of the cross-vectorial product of the stator flux and the stator current space vectors.

$$T_e = \frac{3}{2}P\frac{L_M}{L_2L_1\sigma}\vec{\lambda}_{2dq} \times \vec{\lambda}_{1dq} \tag{5}$$

$$T_e = \frac{3}{2}P\frac{L_M}{L_2L_1\sigma}\left|\vec{\lambda}_{2dq}\right|\left|\vec{\lambda}_{1dq}\right|\sin(\alpha_r - \delta) \tag{6}$$

$$T_e = \frac{3}{2}P\frac{L_M}{L_2L_1\sigma}\left|\vec{\lambda}_{2dq}\right|\left|\vec{\lambda}_{1dq}\right|\sin(\alpha) \tag{7}$$

Tuning PI Regulators for Three-Phase Induction Motor Space Vector Modulation Direct Torque Control
Using Complex Transfer Function Concept

5

Equation (7) shows that variations in stator flux will reflect variations on rotor flux.

Where δ and α_r are the angle of the stator flux and rotor flux space vector with respect to the direct-axis of the synchronous reference frame respectively as is shown in Fig. 1, $\alpha = \alpha_r - \delta$ is the angle between the stator and rotor flux space vectors, P is a number of pole pairs and $\sigma = 1 - L_M^2/(L_1 L_2)$ is the dispersion factor.

Combining equations (1), (2), (3) and (4), after some manipulations, the induction machine model can be written as a complex space state equation in the synchronous reference frame (dq) and the state variables are stator current $\vec{i}_{1dq} = i_{1d} + ji_{1q}$ and stator flux $\vec{\lambda}_{1dq} = \lambda_{1d} + j\lambda_{1q}$ and it is shown in equation (9).

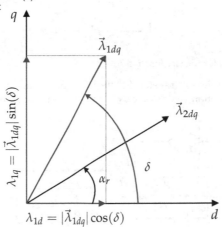

Figure 1. stator and rotor fluxes space vectors in synchronous reference frame.

$$\frac{d\vec{\lambda}_{1dq}}{dt} = -j\omega_1 \vec{\lambda}_{1dq} - R_1 \vec{i}_{1dq} + \vec{v}_{1dq} \tag{8}$$

$$\frac{d\vec{i}_{1dq}}{dt} = a_3 \vec{\lambda}_{1dq} + a_4 \vec{i}_{1dq} + \frac{\vec{v}_{1dq}}{\sigma L_1} \tag{9}$$

$$a_3 = \left(\frac{R_2}{\sigma L_1 L_2} - \frac{jP\omega_{mec}}{\sigma L_1} \right) \tag{10}$$

$$a_4 = -\left[\frac{R_1}{\sigma L_1} + \frac{R_2}{\sigma L_2} + j(\omega_1 - P\omega_{mec}) \right] \tag{11}$$

The machine mechanical dynamics is given by

$$J\frac{d\omega_{mec}}{dt} = \frac{3}{2}P\frac{L_M}{L_2 L_1 \sigma} \vec{\lambda}_{2dq} \times \vec{\lambda}_{1dq} - T_L \tag{12}$$

The ω_1 is the synchronous speed, ω_{mec} is the machine speed, R_1 and R_2 are the estator and rotor windings per phase electrical resistance, L_1 , L_2 and L_m are the proper and mutual inductances of the stator and rotor windings, $\vec{v}$ is the voltage vector , P is the machine number of pair of poles, J is the load and rotor inertia moment, the symbol "*" represents the conjugate of the complex number and T_L is the load torque.

In order to obtain the induction motor complex transfer function the Laplace transform is applied to the equations (8) and (9) in accordance with the complex transfer function concept [24], [10]. Thus, the equation (8) complex transfer function is shown in Figure 2.

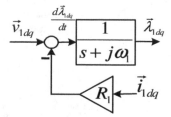

Figure 2. Equation (8) complex transfer function.

And the equation (9) complex transfer function complex transfer function is shown in Figure 3. Thus, the induction motor block diagram originated by use of the equations (8)

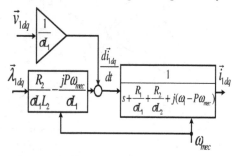

Figure 3. Equation (9) complex transfer function.

and (9) complex transfer functions shown in Figures 2 and 3 and the machine mechanical dynamics (12) is shown in Figure 4. When designing the DTC control system through the

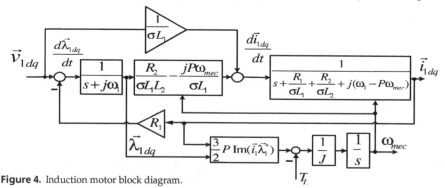

Figure 4. Induction motor block diagram.

Tuning PI Regulators for Three-Phase Induction Motor Space Vector Modulation Direct Torque Control
Using Complex Transfer Function Concept

7

IM complex transfer function, $\vec{v}_{1dq}$ is considered as the input and the $\vec{i}_{1dq}$ is considered as the output. For this purpose it is assumed that the mechanical time constant of the motor is much larger than the transient electromagnetic time constants and the saturation effects is neglected. Thus, ω_{mec} = constant is a valid approximation [24], [11]. Therefore the induction machine complex transfer function $H(s)$ is derived from application of the Laplace transform in equations (8) and (9) and it is the closed loop system of Figure 4 without machine mechanical dynamics. Thus, it has the form given in 13.

$$H(s) = \frac{I_{1dq}}{V_{1dq}} = \frac{\left(\dfrac{s + j\omega_1}{\sigma L_1}\right) + a_3}{(s + j\omega_1)(s + a_4) + R_1 a_3} \tag{13}$$

where $I_{1dq} = \mathcal{L}\left\{\vec{i}_{1dq}\right\}$ and $V_{1dq} = \mathcal{L}\left\{\vec{v}_{1dq}\right\}$.

3. Direct torque control

If the sample time is short enough, such that the stator voltage space vector is imposed to the motor keeping the stator flux constant at the reference value. The rotor flux will become constant because it changes slower than the stator flux.

The electromagnetic torque (14) can be quickly changed by changing the angle α in the desired direction. The angle α can be easily changed when choosing the appropriate stator voltage space vector.

$$T_e = \frac{3}{2}P\frac{L_M}{L_2 L_1 \sigma}\left|\vec{\lambda}_{2\alpha\beta}\right|\left|\vec{\lambda}_{1\alpha\beta}\right|\sin(\alpha) \tag{14}$$

For simplicity, let us assume that the stator phase ohmic drop could be neglected in $\vec{v}_{1\alpha\beta} = R_1\vec{i}_{1\alpha\beta} + \frac{d\vec{\lambda}_{1\alpha\beta}}{dt}$. Therefore $d\vec{\lambda}_{1\alpha\beta}/dt = \vec{v}_{1\alpha\beta}$. During a short time Δt, when the voltage space vector is applied it has:

$$\Delta\vec{\lambda}_{1\alpha\beta} \approx \vec{v}_{1\alpha\beta} \cdot \Delta t \tag{15}$$

Thus, the stator flux space vector moves by $\Delta\vec{\lambda}_{1\alpha\beta}$ in the direction of the stator voltage space vector at a speed which is proportional to the magnitude of the stator voltage space vector. By selecting step-by-step the appropriate stator voltage vector, it is possible to change the stator flux in the required direction.

3.1. Stator flux oriented direct torque control

The stator flux oriented direct torque control (SFO-DTC) have two PI regulators. The outputs of the PI flux and torque controllers can be interpreted as the stator voltage components in the stator flux oriented coordinates as shown in Fig. 5 [23], [3]. The control strategy relies on a simplified description of the stator voltage components, expressed in stator-flux-oriented coordinates as:

$$v_{1d} = R_1 i_{1d} + \frac{d\lambda_1}{dt} \tag{16}$$

$$v_{1q} = R_1 i_{1q} + \omega_1\lambda_1 \tag{17}$$

Where ω_1 is the angular speed of the stator flux vector. The above equations show that the component v_{1d} has influence only on the change of stator flux magnitude, and the component v_{1q}, if the term $\omega_1\lambda_1$ is decoupled, can be used for torque adjustment. Therefore, after coordinate transformation $dq/\alpha\beta$ into the stationary frame, the command values $v_{1d_{ref}}, v_{1q_{ref}},$ are delivered to SVM module.

This SFO-DTC scheme requires the flux and the torque estimators, which can be performed as it is proposed in Fig. 5. Therefore, the control signals are fed to the power electronics drive.

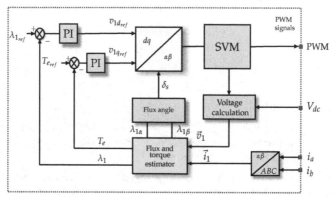

Figure 5. Stator flux oriented direct torque control scheme.

4. Design and tuning PI gains

By using stator field orientation, the torque and stator flux must become parts of a complex number, where the magnitude of the stator flux λ_1 is the real component and the torque T_e is the imaginary component. Hence, the reference signals and the error become a complex number. Thus, the PI regulators presented in the before section [23] has the function to generate a voltage reference space vector using the stator flux-torque error vector $(\varepsilon_\lambda + j\varepsilon_T)$. This way the stator-voltage vector in this control strategy is given by

$$\vec{v}_{1dq_{ref}} = (\varepsilon_\lambda + j\varepsilon_T)\left(Kp + \frac{Ki}{s}\right) \tag{18}$$

Which means that the direct and quadrature axis of the voltage vector are

$$v_{1d_{ref}} = (\varepsilon_\lambda)\left(Kp + \frac{Ki}{s}\right) \tag{19}$$

$$v_{1q_{ref}} = (\varepsilon_T)\left(Kp + \frac{Ki}{s}\right) \tag{20}$$

Where kp is the proportional gain, ki is the integral gain, ε_λ is the flux error signal and ε_T is the torque error signal.

The block diagram of the strategy with the PI regulators is shown in Figure 6.

Tuning PI Regulators for Three-Phase Induction Motor Space Vector Modulation Direct Torque Control
Using Complex Transfer Function Concept

9

The reference stator voltage vector $\vec{v}_{1dq_{ref}}$ is transformed by using stator flux angle δ_s to obtain the stator voltage at stationary reference frame $\alpha\beta$.

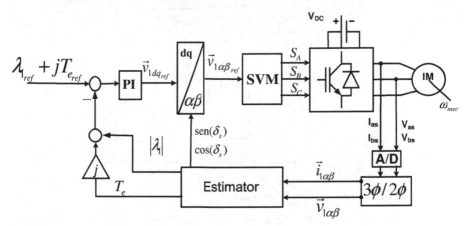

Figure 6. DTC strategy with PI regulators and complex signals.

4.1. Stator flux estimation

The stator flux estimation is done by

$$\vec{\lambda}_{1\alpha\beta} = \int (\vec{v}_{1\alpha\beta} - R_1\vec{i}_{1\alpha\beta})dt \tag{21}$$

A satisfactory flux estimation for induction motor at low speed using Equation (21) is obtained by using the integration method presented in [8] and the block diagram for the flux stimation is presented in Figure 7

The stator flux angle is estimated by using the trigonometric transfer function

$$\delta_s = \arctan\left(\frac{\lambda_{1\beta}}{\lambda_{1\alpha}}\right) \tag{22}$$

4.2. Design of the PI regulator gains

In order to tune the PI regulator it is necessary the closed-loop complex transfer function of the controlled induction motor. The complex transfer function of the controlled induction motor was also used to tune a complex gain controller in which has been presented in [7].

In accordance with the DTC control strategy the induction motor output has to be the stator flux magnitude λ_1 and the torque T_e. Therefore the H(s) (13) outputs have to become the stator flux magnitude λ_1 and the torque T_e. The expression to obtain the stator flux by using the stator current i_{1d} is given by

$$\lambda_1 = \lambda_{1d} \cong G\sigma L_1 i_{1d} \tag{23}$$

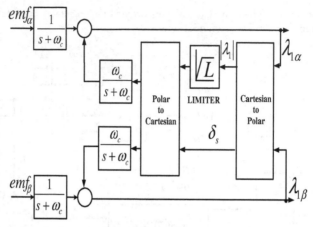

Figure 7. Block diagram for the stator flux stimation.

and to obtain the electromagnetic torque in the dq reference frame one may use the expression:

$$T_e = \frac{3}{2} P \lambda_1 i_{1q} \tag{24}$$

As the stator flux magnitude λ_1 is assumed to be essentially constant through of the equations (13), (23) and (24) the new transfer function is achieved with torque and flux as output and it is given by

$$\frac{X_{\lambda T}}{V_{1dq}} = H(s) \left(G\sigma L_1 + jP\frac{3}{2}\lambda_1 \right) \tag{25}$$

where $X_{\lambda T} = \mathcal{L}\{\lambda_1 + jT_e\}$.

The low speeds utilized in this book chapter are 60 rpm ($6.25rad/s$), 125 rpm ($13rad/s$), 150 rpm ($16rad/s$), 180 rpm ($17rad/s$) that corresponds to 2Hz, 4.16Hz, 5Hz and 6Hz respectively. The frequency-response function of Equation (25) is presented in Figure 8 at frequencies 2Hz, 4.16Hz, 5Hz and 6Hz in accordance with the induction motor desired speed.

Then, from Equations (18) and (25) one obtain the control system block diagram and it is shown in Figure 9.

The expression of the closed loop transfer function of the system to design the PI regulators showed in Figure 9 is given by

$$\frac{X_{\lambda T}}{X_{\lambda T_{ref}}} = \frac{\left(Kp + \frac{Ki}{s} \right) H(s) \left(G\sigma L_1 + jP\frac{3}{2}\lambda_1 \right)}{1 + \left(Kp + \frac{Ki}{s} \right) H(s) \left(G\sigma L_1 + jP\frac{3}{2}\lambda_1 \right)} \tag{26}$$

where $X_{\lambda T_{ref}} = \mathcal{L}\left\{ \lambda_{1ref} + jT_{eref} \right\}$.

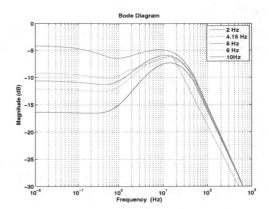

Figure 8. Equation (25) frequency-response function.

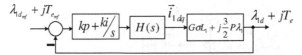

Figure 9. System to design the PI regulator.

As the variables at synchronous referential are constants the angle of output $X_{\lambda T}$ is neglected. At the frequency of 2Hz, 4.16Hz and 6Hz the kp and ki gains are chosen by using simulations, considering slip approximately null and the 0 dB magnitude. Them values are $kp = 155$ and $ki = 15$. The frequency-response function of Equation (26) is shown in Figure 10 and its magnitude is near 0 dB.

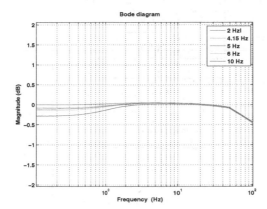

Figure 10. Frequency-response function of the equation (26)

5. Experimental results

The DTC strategy were implemented using a Texas Instruments DSP TMS320F2812 platform. The system consists of a three-phase voltage source inverter with insulated-gate bipolar transistors (IGBTs) and the three-phase induction motor parameters are shown in the appendix. The stator voltage commands are modulated by using symmetrical space vector PWM, with switching frequency equal to 2.5 kHz. The DC bus voltage of the inverter is 226 V. The stator voltages and currents are sampled in the frequency of 2.5 kHz. A conventional PI regulators generates a torque reference by using the speed error. The flux and torque estimation, and the flux and torque PIs regulators and speed controller have the same sampling frequency of 2.5 kHz. The encoder resolution is 1500 pulses per revolution. The algorithm of the DTC strategy was programmed on the Event Manager 1 of the Texas Instruments DSP TMS320F2812 platform and its flowchart is presented in Figure 11 and the schematic of implemention is presented in Figure 12.

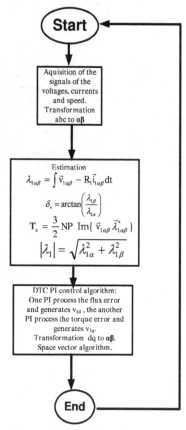

Figure 11. The flowchart of the DSP program.

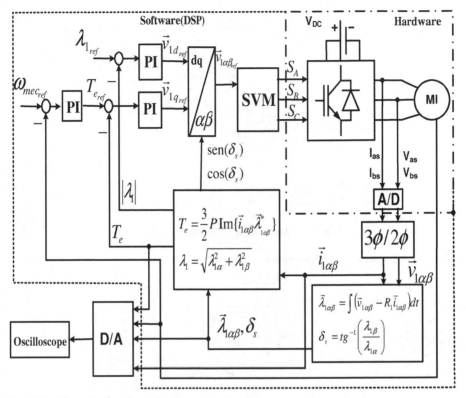

Figure 12. Schematic of implemention.

Five no-load induction motor tests were made. The first one was the response to a torque step of 12.2 Nm which is shown in Figure 13. It can be seen the satisfactory response of torque although it has oscillation. This oscillation occurs due to the natural lack of accuracy in the measurements of currents, voltages and parameters variations.

Figure 14 shows when the speed varies from 6.28 rad/s to 18.85 rad/s in 200 ms. This result confirms the satisfactory performance of the controller due to the fact that the the speed reaches the reference in several conditions although the gains of PI are designed for induction motor speed operation at 2 Hz and 6 Hz.

In the third test the speed varies in forward and reversal operation and the result are presented in Figures 15(a) and 15(b). The speed changes from 13 rad/s to -13 rad/s in 1 s and the gains of PI regulator are not changed during the test. This result confirms the satisfactory performance of the controller due to the fact that the the speed reaches the reference in several conditions and the PI regulator was designed for induction motor speed operation at 4.15 Hz. The small error occurs due the natural lack of accuracy in the measurement of the speed.

Figure 16 presents the speed response when the speed varies from 6.28 rad/s to -6.28 rad/s. The result confirms the satisfactory performance of the PI regulator again due to the fact that

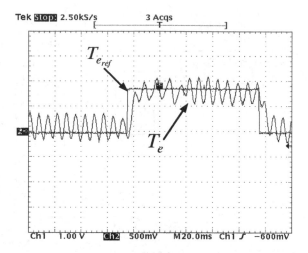

Figure 13. Responses to step torque operation (9 Nm/div).

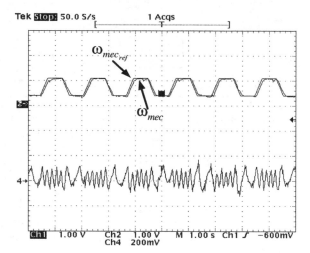

Figure 14. Speed forward and reversal operation (15.7 ras/s.div) and a phase current (10 A/div).

the speed reaches the reference value and the gains of PI are designed for induction motor speed operation at 2 Hz.

In load test the speed reference was 36.6 rad/s and a load torque of 11.25N.m was applied to the motor. In this test a dc generator is coupled to the rotor of induction motor. So the generated voltage of the DC generator is conected to the load with variable resistance. The test is shown in Figure 17 and the steady state error is 4.5%.

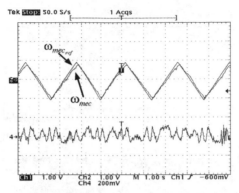

(a) Speed reversal operation (12.57 rad/s.div).

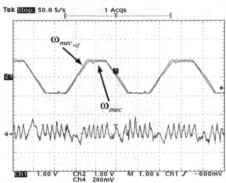

(b) Speed forward and reversal (13 rad/s.div).

Figure 15. Speed forward and reversal operation and *a* phase current (10 A/div)

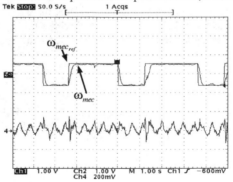

Figure 16. Speed response to step operation (12.57 rad/s.div) and *a* phase current (10 A/div).

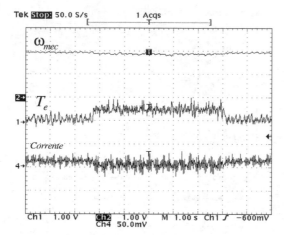

Figure 17. Load test (18,3 rad/s.div) and *a* phase current (20 A/div).

6. Conclusion

In this book chapter was presented a method to design and tune the PI regulators for the three-phase IM DTC-SVM strategy using the mathematical model complex transfer function when the machine operates at low speed. The concept of complex transfer function allows to obtain the PI regulator gains by using the closed loop system frequency response function of the controlled induction motor.

The experimental results shown the satisfactory performance of the regulator due to the fact that the speed reaches the reference value in several conditions although the complex gain was designed for a limited points of induction motor operation. Thus, the design of PI regulator has an acceptable performance although an detailed analysis considering parameters variations and other several speed operations has to be done. Due to the variable speed operation maybe it will be necessary to construct a table with PI gains designed for each desired speed or to an each speed range. The PI regulator overcomes the low speed operations shortcomings to the IM DTC-SVM strategy with a minor complexity. Thus, the complex transfer function becomes an interesting tool for design and tune PI regulator for IM drives.

Appendix

Three-phase induction motor variables and parameters: $PN = 2.3kW; V_N = 220\ V; Poles = 4$ $R_1 = 2.229\ \Omega; R_2 = 1.522\ \Omega; L_m = 0.238485\ H; L_1 = 0.2470\ H; L_2 = 0.2497\ H; J = 0.0067\ Kgm^2$.

Acknowledgment

The authors are grateful to CAPES, CNPq and FAPESP for the financial support for this research.

Author details

Alfeu J. Sguarezi Filho
Universidade Federal do ABC, Brazil

José Luis Azcue and Ernesto Ruppert
School of Electrical and Computer Engineering, University of Campinas, Brazil

7. References

[1] Blaschke, F. [1977]. The principle of field orientation control as applied to the new transvector closed loop control system for rotating machines, *Siemens Review* 39(5): 217–220.

[2] Briz, F., degener, M. W. & Lorenz, R. D. [2000]. Analysis and design of current regulators using complex vectors, *IEEE Trans. Ind. Applicat.* 32: 817–825.

[3] Buja, G. & Kazmierkowski, M. [2004]. Direct torque control of pwm inverter-fed ac motors - a survey, *Industrial Electronics, IEEE Transactions on* 51(4): 744–757.

[4] Cad, M. M. & de Aguiar, M. L. [2000]. The concept of complex transfer functions applied to the modeling of induction motors, *IEEE Winter Meeting 2000 of the IEEE Power Engineering Society* .

[5] Casadei, D., Serra, G. & Tani, A. [2001]. Steady-state and transient performance evaluation of a dtc scheme in the low speed range, *IEEE Trans. on Power Electronics* 16(6): 846–851.

[6] Depenbrock, M. [1988]. Direct self-control(dsc) of inverter-fed induction machine, *IEEE Trans. Power Electronics* 3(4): 420–429.

[7] Filho, A. J. S. & Filho, E. R. [2008]. The complex controller applied to the induction motor control, *IEEE Applied Power Electronics Conference and Exposition - APEC* pp. 1791–1795.

[8] Filho, A. J. S. & Filho, E. R. [2009]. The complex controller for three-phase induction motor direct torque control, *Sba Controle e Automação* 20(2).

[9] Gataric, S. & Garrigan, N. R. [1999]. Modeling and design of three-phase systems using complex transfer functions, *IEEE Trans. Ind. Electron.* 42: 263–271.

[10] Holtz, J. [1995]. The representation of ac machine dynamics by complex signal flow graphs, *IEEE Trans. Ind. Electron.* 42: 263–271.

[11] Holtz, J., Quan, J., Pontt, J., Rodríguez, J., newman, P. & Miranda, H. [2004]. Design of fast and robust current regulators for high-power drives based on complex state variables, *IEEE Trans. Ind. Applications* 40: 1388–1397.

[12] Kenny, B. H. & Lorenz, R. D. [2001]. Stator and rotor flux based deadbeat direct torque control ofinduction machines, *IEEE Industry Applications Conference* 1: 133–139.

[13] Kovács, P. K. & Rácz, E. [1984]. *Transient Phenomena in Electrical Machines*, Amsterdam, The Netherlands: Elsevier.

[14] Kumsuwana, Y., Premrudeepreechacharna, S. & Toliyat, H. A. [2008]. Modified direct torque control method for induction motor drives based on amplitude and angle control of stator flux, *Electric Power Systems Research* 78: 1712–1718.

[15] Lee, K.-B., Blaabjerg, F. & Yoon, T.-W. [2007]. Speed-sensorless dtc-svm for matrix converter drives with simple nonlinearity compensation, *IEEE Transactions on Industry Applications* 43(6): 1639–1649.

[16] Leonhard, W. [1985]. *Control of Electrical Drives*, Springer-Verlag Berlin Heidelberg New York Tokyo.

[17] Novotny, D. W. & Lipo, T. A. [1996]. *Vector Control and Dynamics of AC Drives*, Clarendon Press OXFORD.

[18] Phillips, C. [2000]. *Feedback Control Systems*, Pretince Hall.

[19] Ryu, J. H., Lee, K. W. & Lee, J. S. [2006]. A unified flux and torque control method for dtc-based induction-motor drives, *IEEE Trans. on Power Electronics* 21(1): 234–242.

[20] Shady M. Gadoue, D. G. & Finch, J. W. [2005]. Tuning of pi speed controller in dtc of induction motor based on genetic algorithms and fuzzy logic schemes, *International Conference on Technology and Automation* .

[21] Sheu, T.-T. & Chen, T.-C. [1999]. Self-tuning control of induction motor drive using neural network identifier, *IEEE Transactions on Energy Conversion* 14(4).

[22] Takahashi, I. & Noguchi, T. [1986]. A new quick-response and high-efficiency control strategy of an induction motor, *Industry Applications, IEEE Transactions on* IA-22(5): 820 –827.

[23] Xue, Y., Xu, X., Habetler, T. G. & Divan, D. M. [1990]. A low cost stator flux oriented voltage source variable speed drive, *Conference Record of the 1990 IEEE Industrial Aplications Society Annual Meetting* 1: 410–415.

[24] Yamamura, S. [1992]. *Spiral Vector Theory of AC Circuits and Machines*, Clarendon Press OXFORD.

Chapter 2

Advanced Control Techniques for Induction Motors

Manuel A. Duarte-Mermoud and Juan C. Travieso-Torres

Additional information is available at the end of the chapter

1. Introduction

The design of suitable control algorithms for induction motors (IM) has been widely investigated for more than two decades. Since the beginning of field oriented control (FOC) of AC drives, seen as a viable replacement of the traditional DC drives, several techniques from linear control theory have been used in the different control loops of the FOC scheme, such as Proportional Integral (PI) regulators, and exact feedback linearization (Bose, 1997, 2002; Vas, 1998). Due to their linear characteristics, these techniques do not guarantee suitable machine operation for the whole operation range, and do not consider the parameter variations of the motor-load set.

Several nonlinear control techniques have also been proposed to overcome the problems mentioned above, such as sliding mode techniques (Williams & Green, 1991; Al-Nimma & Williams, 1980; Araujo & Freitas, 2000) and artificial intelligence techniques using fuzzy logic, neuronal networks or a combination of them (Vas, 1999; Al-Nimma & Williams, 1980; Bose, 2002). All these techniques are based on complex control strategies differing of the advanced control techniques described here.

In this chapter we present a collection of advanced control strategies for induction motors, developed by the authors during the last ten years, which overcome some of the disadvantages of the previously mentioned control techniques. The techniques studied and presented in this chapter are based on equivalent passivity by adaptive feedback, passivity by interconnection and damping assignment (IDA-PCB) and fractional order proportional-integral controller (FOPIC) in the standard field oriented control scheme (FOC).

All of the control strategies described here guarantee high performance control, such as high starting torque at low speed and during the transient period, accuracy in steady state, a wide range of speed control, and good response under speed and load changes. For all of the control strategies developed throughout the chapter, after a brief theoretical description of each one of them, simulation as well as experimental results of their application to control IM are presented and discussed.

The main contribution of this Chapter is to show that IM control techniques based on passivity, IDA-PCB and FOPIC can be successfully used in a FOC scheme, presenting some advantages over the classical techniques.

2. Adaptive passivity based control for the IM

Four novel adaptive passivity based control (APBC) techniques were first developed by the main author and his collaborators. As explained in Sections 2.1 and 2.2, these are the adaptive approach of feedback passive equivalence controllers, which were developed for SISO (Castro-Linares & Duarte-Mermoud, 1998; Duarte-Mermoud et al, 2001) and MIMO systems (Duarte-Mermoud et al 2003; Duarte-Mermoud et al, 2002), including controllers with fixed adaptive gains (CFAG) as well as controllers with time-varying adaptive gains (CTVAG). The nonlinear model characteristics were considered in the controller design and they are adaptive in nature, guaranteeing robustness under all model parameter variations.

These techniques were developed for systems parameterized in the so called normal form with explicit linear parametric dependence, which are also locally weakly minimum phase. It can be verified that the IM can be expressed in that particular form and therefore these strategies can be readily applied to them.

Based on the APBC control techniques developed for SISO systems, previously presented, two novel control strategies for induction motors were proposed in Travieso-Torres & Duarte-Mermoud (2008) as described in Section 2.3. Besides, a MIMO version, based on the MIMO techniques already mentioned, was applied to the IM in Duarte-Mermoud & Travieso-Torres (2003) and described here in Section 2.4. Results from SISO and MIMO controllers are similar, however the SISO controllers present only two adjustable parameters by means of simple adaptive laws, being simpler than the solution for the MIMO case, since the MIMO controllers have a larger number of adjustable parameters.

These controllers are applied to the IM considering the scheme presented in Figure 1. In both cases, SISO and MIMO controllers were suitably simplified using the Principle of Torque-Flux Control (PTFC) proposed in Travieso (2002). This principle is applicable to strategies working under a FOC scheme. Based on the PTFC, the design of the SISO and MIMO controllers do not require flux estimations.

For the SISO and the MIMO approaches, the proposed CFAG is simpler, but better transient behaviour was obtained when CTVAG was used. The results were compared with the classical basic control scheme (BCS) shown in Figure 2 (Chee-Mun, 1998), concluding that the proposed adaptive controllers showed a better transient behaviour. In addition, CFAG and CTVAG do not need the knowledge of the set motor–load parameters and robustness under variations of such parameters is guaranteed.

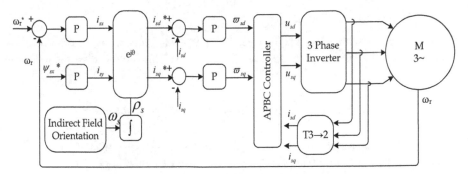

Figure 1. Proposed control scheme with field oriented block (APBC)

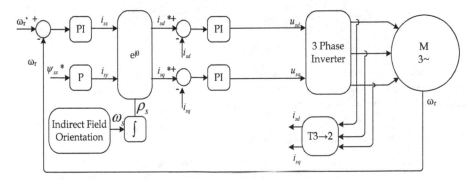

Figure 2. Basic control scheme with field oriented block (BCS)

2.1. SISO Adaptive Passivity Based Control (ABPC) theory

The SISO APBC approach was proposed in Castro-Linares & Duarte-Mermoud (1998) and Duarte-Mermoud et al (2001), for systems parameterized in the following normal form (Byrnes et al, 1991), with explicit linear parametric dependence

$$\dot{y} = \Lambda_a^T A(y,z) + \Lambda_b B(y,z) u$$
$$\dot{z} = \Lambda_0 f_o(z) + \Lambda_p P(y,z) y \tag{1}$$

with $z \in \mathfrak{R}^n$, $y \in \mathfrak{R}$, $u \in \mathfrak{R}$, $A(y, z) \in \mathfrak{R}^m$, $B(y, z) \in \mathfrak{R}$, $f_0 \in \mathfrak{R}^n$, $P(y, z) \in \mathfrak{R}^n$; and the parameters $\Lambda_a \in \mathfrak{R}^m$, $\Lambda_b \in \mathfrak{R}$, $\Lambda_0 \in \mathfrak{R}^{n \times n}$, $\Lambda_p \in \mathfrak{R}^{n \times n}$. The function $\dot{z} = \Lambda_0 f_0(z)$ is known as zero dynamics (Isidori, 1995; Nijmiejer & Van der Shaft, 1996). Besides, it is necessary to check that the system is locally weakly minimum phase by finding a positive definite differentiable function $W_0(z)$ satisfying $\left(\partial W_0(z) / \partial z\right)^T \Lambda_0 f_0(z) \leq 0, \forall \Lambda_0$ (Byrnes et al, 1991). According to the theory presented in the original papers, for locally weakly minimum phase systems of the form (1) with matrix $B(y,z)$ being invertible, there exist two adaptive controllers guaranteeing stability described in the following section.

2.1.1. SISO controller with fixed gains

A SISO controller with fixed adaptive gains (CFAG) was proposed in Castro-Linares & Duarte-Mermoud (1998) for SISO systems of the form (1). This controller has the following form

$$u(y,z,\theta_h) = \frac{1}{B}\left[\theta_1^T(t)A(y,z) - \theta_2(t)P(y,z)\frac{\partial W_0(z)}{\partial z} + \theta_3(t)\varpi\right] \tag{2}$$

with $z \in \mathfrak{R}^2$, $y \in \mathfrak{R}$, $u \in \mathfrak{R}$, $A(y, z) \in \mathfrak{R}$, $B(y, z) \in \mathfrak{R}$, $f_0 \in \mathfrak{R}^2$, $P(y, z) \in \mathfrak{R}^2$ and the adjustable parameters $\theta_1(t) \in \mathfrak{R}^p$ and $\theta_2(t), \theta_4(t) \in \mathfrak{R}$ updated with the adaptive laws

$$\dot{\theta}_1(t) = -sign(\Lambda_b)A(y,z)y$$
$$\dot{\theta}_2(t) = -sign(\Lambda_b)yP(y,z)\left(\frac{\partial W_0(z)}{\partial z}\right) \tag{3}$$
$$\dot{\theta}_3(t) = -sign(\Lambda_b)y\varpi$$

that applied to system (1) make it locally feedback equivalent to a C²-passive system from the new input ϖ to the output y. The parameters $\Lambda_a \in \mathfrak{R}$, $\Lambda_b \in \mathfrak{R}$, $\Lambda_0 \in \mathfrak{R}$, $\Lambda_p \in \mathfrak{R}^{2x2}$ represent constant but unknown parameters from a bounded compact set Ω.

2.1.2. SISO controller with time-varying gains

Another adaptive controller approach but with time-varying gains (Duarte-Mermoud et al, 2001) was also proposed for a SISO system of the form (1). This controller has the same control law shown in (2), but updated with the following adaptive laws

$$\dot{\theta}_1(t) = -sign(\Lambda_b)\left(\gamma_1^{-1}(t)/\sqrt{1+\frac{1}{\gamma(t)^T\gamma(t)}}\right)A(y,z)y$$

$$\dot{\theta}_2(t) = -sign(\Lambda_b)\left(\gamma_2^{-1}(t)/\sqrt{1+\frac{1}{\gamma(t)^T\gamma(t)}}\right)P(y,z)\left(\frac{\partial W_0(z)}{\partial z}\right)y \tag{4}$$

$$\dot{\theta}_3(t) = -sign(\Lambda_b)\left(\gamma_3^{-1}(t)/\sqrt{1+\frac{1}{\gamma(t)^T\gamma(t)}}\right)\varpi y$$

where $\gamma_1(t) \in \mathfrak{R}^{4\times4}$ and $\gamma_2(t), \gamma_3(t), \gamma_4(t) \in \mathfrak{R}$ are time-varying adaptive gains defined by

$$\dot{\gamma}_1(t) = -\left[\gamma_1 A(y,z)A^T(y,z)\gamma_1\right],$$
$$\dot{\gamma}_2(t) = \left[\gamma_2(t)P(y,z)\left(\frac{\partial W_0(z)}{\partial z}\right)\right]^2, \qquad \gamma(t) = \left[Trace\{\gamma_1(t)\} \quad \gamma_2(t) \quad \gamma_3(t)\right] \in \mathfrak{R}^3 \tag{5}$$
$$\dot{\gamma}_3(t) = \left[\gamma_3(t)\varpi\right]^2.$$

2.2. MIMO Adaptive Passivity Based Control (ABPC) theory

The MIMO APBC approach was proposed in Duarte-Mermoud et al (2002) and Duarte-Mermoud et al (2003), for systems parameterized in the following normal form (Byrne et al, 1991), with explicit linear parametric dependence

$$
\begin{aligned}
\dot{y} &= \Lambda_a A(y,z) + \Lambda_b B(y,z)u \\
\dot{z} &= \Lambda_0 f_0(z) + P^T(y,z)\Lambda_p y
\end{aligned}
\tag{6}
$$

with $z \in \Re^2$, $y \in \Re^2$, $u \in \Re^2$, $A(y,z) \in \Re^8$, $B(y,z) \in \Re^{2x2}$, $f_0 \in \Re^2$, $P(y,z) \in \Re^{2x2}$. The parameters $\Lambda_a \in \Re^{2x8}$, $\Lambda_b \in \Re^{2x2}$, $\Lambda_0 \in \Re^{2x2}$ and $\Lambda_p \in \Re^{2x2}$ represent constant but unknown parameters from a bounded compact set Ω. The term $\dot{z} = \Lambda_0 f_0(z)$ is the so called zero dynamics (Isidori, 1995; Nijmiejer & Van der Shaft, 1996). In this case it is also necessary to check that system (6) is locally weakly minimum phase by finding a positive definite differentiable function $W_0(z)$ satisfying $\left(\partial W_0(z)/\partial z\right)^T \Lambda_0 f_0(z) \leq 0, \forall \Lambda_0$ (Byrnes et al, 1991). According to the theory presented in the original papers, for locally weakly minimum phase systems of the form (11) with matrix $B(y,z)$ being invertible, there exist two type of adaptive controllers guaranteeing stability which are described in the following section.

2.2.1. MIMO controller with fixed gains

According to Duarte-Mermoud et al (2002) there exists an adaptive controller of the form

$$
u(t) = \left[\theta_1(t)A(y,z) - \theta_2(t)P(y,z)\frac{\partial W_0(z)}{\partial z} + \theta_3(t)\varpi(t) \right]
\tag{7}
$$

with the adaptive laws

$$
\begin{aligned}
\dot{\theta}_1(t) &= -yA^T(y,z) \\
\dot{\theta}_2(t) &= -y\left(\frac{\partial W_0(z)}{\partial z}\right)^T P^T(y,z) \\
\dot{\theta}_3(t) &= -y\varpi^T(t)
\end{aligned}
\tag{8}
$$

that applied to system (6) make it locally feedback equivalent to a C^2-passive system from the input $\varpi(t)$ to the output $y(t)$. The parameters $\theta_1(t) \in \Re^{2x8}$, $\theta_2(t) \in \Re^{2x2}$ and $\theta_3(t) \in \Re^{2x2}$ represent adjustable controller parameters whose ideal values are $\theta_1 = -\Lambda_b^{-1}\Lambda_a \in \Re^{2x8}$, $\theta_2 = -\Lambda_b^{-1}\Lambda_p^T \in \Re^{2x2}$ and $\theta_3 = \Lambda_b^{-1} \in \Re^{2x2}$.

2.2.2. MIMO controller with time-varying gains

On the other hand, CTVAG was proposed in Duarte-Mermoud et al (2003). This controller has the same form (7), but with adaptive laws given by

$$\dot{\theta}_1(t) = -yA^T(y,z)\frac{\Gamma_1^{-1}}{\sqrt{1 + Trace\left(\Gamma_1^{-2} + \Gamma_2^{-2} + \Gamma_3^{-2}\right)}}$$

$$\dot{\theta}_2(t) = -\frac{\Gamma_2^{-1}}{\sqrt{1 + Trace\left(\Gamma_1^{-2} + \Gamma_2^{-2} + \Gamma_3^{-2}\right)}} y\left(\frac{\partial W_0(z)}{\partial z_1}\right)^T P^T(y,z) \tag{9}$$

$$\dot{\theta}_3(t) = -\frac{\Gamma_3^{-1}}{\sqrt{1 + Trace\left(\Gamma_1^{-2} + \Gamma_2^{-2} + \Gamma_3^{-2}\right)}} y\varpi^T(t)$$

and time-varying adaptive gains defined by

$$\dot{\Gamma}_1 = -\Gamma_1 A(y,z)A^T(y,z)\Gamma_1, \qquad\qquad \Gamma_1(t_0) > 0$$

$$\dot{\Gamma}_2 = -\Gamma_2 P(y,z)\left(\frac{\partial W_0(z)}{\partial z}\right)\left(\frac{\partial W_0(z)}{\partial z}\right)^T P^T(y,z)\Gamma_2, \qquad \Gamma_2(t_0) > 0 \tag{10}$$

$$\dot{\Gamma}_3 = -\Gamma_3 \varpi(t)\varpi^T(t)\Gamma_3, \qquad\qquad \Gamma_3(t_0) > 0$$

According to Duarte-Mermoud et al (2002), this controller applied to system (6) will convert it to an equivalent C^2-passive system from the input $\varpi(t)$ to the output $y(t)$. The parameters $\theta_1(t) \in \mathfrak{R}^{2 \times 8}$, $\theta_2(t) \in \mathfrak{R}^{2 \times 2}$ and $\theta_3(t) \in \mathfrak{R}^{2 \times 2}$ represent adjustable controller parameters whose ideal values are $\theta_1 = -\Lambda_b^{-1}\Lambda_a \in \mathfrak{R}^{2 \times 8}$, $\theta_2 = -\Lambda_b^{-1}\Lambda_p^T \in \mathfrak{R}^{2 \times 2}$ and $\theta_3 = \Lambda_b^{-1} \in \mathfrak{R}^{2 \times 2}$

2.3. SISO ABPC applied to the IM

In this Section the design of SISO CFAG and SISO CTVAG for IM is explained, based on the SISO theories previously described.

2.3.1. SISO IM modeling

In order to apply the controllers described in Section 2.1 the IM model was expressed as SISO subsystems parameterized in the following locally weakly minimum phase normal form with explicit linear parametric dependence. For Subsystem 1 we have

$$\Lambda_{a1} = \left[-\frac{R'_s}{\sigma L_s} \quad 1 \quad \frac{L_m R_r}{\sigma L_s L_r^2} \quad \frac{L_m}{\sigma L_s L_r}\right]^T, \quad A_1(y_i,z) = \begin{bmatrix} e_{i_{sx}} \\ \omega_g e_{i_{sy}} \\ e_{\psi_{rx}} \\ \omega_r e_{\psi_{ry}} \end{bmatrix}, \quad \Lambda_{b1} = \frac{1}{\sigma L_s}, \quad B_1(y_i,z) = 1,$$

$$\Lambda_{p1} = \begin{bmatrix} \dfrac{L_m}{T_r} & 0 \\ 0 & \dfrac{L_m}{T_r} \end{bmatrix}, \quad y_1 = e_{i_{sx}}, \quad P_1(y_i,z) = \begin{bmatrix} P_{11} \\ P_{21} \end{bmatrix} = \begin{bmatrix} 1 \\ \dot{e}_{i_{sy}} \\ \dot{e}_{i_{sx}} \end{bmatrix}, \quad u_1 = e_{u_{sx}}. \tag{11}$$

For Subsystem 2 we can write

$$\Lambda_{a2} = \left[-\frac{R'_s}{\sigma L_s} \quad -1 \quad \frac{L_m R_r}{\sigma L_s L_r^2} \quad -\frac{L_m}{\sigma L_s L_r} \right]^T, \quad A_2(y_i, z) = \begin{bmatrix} e_{i_{sy}} \\ \omega_g e_{i_{sx}} \\ e_{\psi_{ry}} \\ \omega_r e_{\psi_{rx}} \end{bmatrix}, \quad \Lambda_{b2} = \frac{1}{\sigma L_s}, \quad B_2(y_i, z) = 1,$$

(12)

$$\Lambda_{p2} = \begin{bmatrix} \dfrac{L_m}{T_r} & 0 \\ 0 & \dfrac{L_m}{T_r} \end{bmatrix}, \quad y_2 = e_{i_{sy}}, \quad P_2(y_i, z) = \begin{bmatrix} P_{12} \\ P_{22} \end{bmatrix} = \begin{bmatrix} \dot{e}_{i_{sx}} \\ \dot{e}_{i_{sy}} \\ 1 \end{bmatrix}, \quad u_2 = e_{u_{sy}}.$$

2.3.2. Principle of torque – Flux control

The PTFC, proposed in Travieso (2002), states that in controlling the torque and flux for IM, the controllers design can be focused only to control the stator currents. This is true for the case when a scheme with coordinate transformation block $e^{j\rho_g}$ (Field Oriented Scheme), to transform from a stationary to a rotating coordinate system, is considered. Therefore, it is pointless to make efforts to directly control rotor flux or rotor current components. It is proven in Travieso (2002) that the controller still guarantees suitable control of the torque and flux and making it possible to discard all the terms concerning the rotor current or rotor flux components in its design.

2.3.3. SISO CFAG applied to the IM

In Travieso-Torres & Duarte-Mermoud (2008) a simplified controller for IM was proposed based on the theories from Castro-Linares & Duarte-Mermoud (1998). After applying the PTFC and considering the controller directly feeding the IM in the stator coordinate system, this means that $\omega g = 0$, this SISO controller has the following form

$$\left. \begin{aligned} u_i(y_i, z, \theta_{hi}) &= \theta_{1i} y_i + \theta_{4i} \varpi_i \\ \dot{\theta}_{1i} &= -y_i^2 \\ \dot{\theta}_{4i} &= -y_i \varpi_i \end{aligned} \right\} \quad \begin{aligned} i &= 1, 2 \\ h &= 1, 4 \end{aligned}$$

(13)

2.3.4. SISO CTVAG applied to the IM

Another adaptive controller but with time-varying gains was also proposed in Travieso-Torres & Duarte-Mermoud (2008), based on the results of Duarte-Mermoud & Castro-Linares (2001). This controller, after applying the PTFC and considering the controller directly feeding the motor in the stator coordinate system, has the following form

$$u_i(y_i, z, \theta_{hi}) = \theta_{1i} y_i + \theta_{4i} \varpi_i$$

$$\dot{\theta}_{1i} = -sign(\Lambda_{bi}^*) \left(\gamma_1^{-1} / \sqrt{1 + \frac{1}{\gamma_i^T \gamma_i}} \right) y_i^2 ,$$

$$\dot{\theta}_{4i} = -sign(\Lambda_{bi}^*) \left(\gamma_4^{-1} / \sqrt{1 + \frac{1}{\gamma_i^T \gamma_i}} \right) \varpi_i y_i ,$$

with
$$\dot{\gamma}_{1i} = -(\gamma_{1i} y_i)^2$$
and
$$\dot{\gamma}_{4i} = -(\gamma_{4i} \varpi_i)^2$$

$$\begin{matrix} i = 1,2 \\ h = 1,4 \end{matrix} \quad (14)$$

2.4. MIMO ABPC applied to the IM

In this Section the design of the MIMO CFAG and the MIMO CTVAG for the IM are presented, based on the MIMO theories previously stated.

2.4.1. MIMO model of the IM

In order to apply controllers from Duarte-Mermoud et al (2002) and Duarte-Mermoud et al (2003), the IM model was expressed in form (6) as follows

$$\dot{y} = \begin{bmatrix} -\dfrac{R'_s}{\sigma L_s} & 0 & 0 & 1 & \dfrac{L_m R_r}{\sigma L_s L_r^2} & 0 & 0 & -\dfrac{L_m}{\sigma L_s L_r} \\ 0 & -\dfrac{R'_s}{\sigma L_s} & -1 & 0 & 0 & \dfrac{L_m R_r}{\sigma L_s L_r^2} & \dfrac{L_m}{\sigma L_s L_r} & 0 \end{bmatrix} \begin{bmatrix} e_{i_{sx}} \\ e_{i_{sy}} \\ \omega_g e_{i_{sx}} \\ \omega_g e_{i_{sy}} \\ e_{\psi_{rx}} \\ e_{\psi_{ry}} \\ \omega_r e_{\psi_{rx}} \\ \omega_r e_{\psi_{ry}} \end{bmatrix} + \begin{bmatrix} \dfrac{1}{\sigma L_s} & 0 \\ 0 & \dfrac{1}{\sigma L_s} \end{bmatrix} I_2 u,$$

$$\dot{z} = \begin{bmatrix} -\dfrac{R_r}{L_r} & (\omega_g - \omega_r) \\ -(\omega_g - \omega_r) & -\dfrac{R_r}{L_r} \end{bmatrix} \begin{bmatrix} e_{\psi_{rx}} \\ e_{\psi_{ry}} \end{bmatrix} + I_2 \begin{bmatrix} \dfrac{L_m}{T_r} & 0 \\ 0 & \dfrac{L_m}{T_r} \end{bmatrix} y, \qquad (15)$$

with $z = \begin{bmatrix} e_{\psi_{rx}} \\ e_{\psi_{ry}} \end{bmatrix}$, $y = \begin{bmatrix} e_{i_{sx}} \\ e_{i_{sy}} \end{bmatrix}$, $u = \begin{bmatrix} e_{u_{sx}} \\ e_{u_{sy}} \end{bmatrix}$

with $T_r = L_r/R_r$

2.4.2 MIMO CFAG applied to the IM

According to Duarte-Mermoud and Travieso-Torres (2003) there exist an adaptive controller of the form

$$u(t) = \theta_1(t)y^T + \theta_3(t)\varpi(t)$$

$$\dot{\theta}_1(t) = -yy^T \qquad (16)$$

$$\dot{\theta}_3(t) = -y\varpi^T(t)$$

that applied to system (6) makes it locally feedback equivalent to a C^2-passive system from the input $\varpi(t)$ to the output $y(t)$. The parameters $\theta_1(t) \in \mathfrak{R}^{2\times8}$, $\theta_2(t) \in \mathfrak{R}^{2\times2}$ and $\theta_3(t) \in \mathfrak{R}^{2\times2}$ represent adjustable controller parameters whose ideal values are $\theta_1 = -\Lambda_b^{-1}\Lambda_a \in \mathfrak{R}^{2\times8}$, $\theta_2 = -\Lambda_b^{-1}\Lambda_p^T \in \mathfrak{R}^{2\times2}$ and $\theta_3 = \Lambda_b^{-1} \in \mathfrak{R}^{2\times2}$.

2.4.3. MIMO CTVAG applied to the IM

Finally a CTVAG was proposed in Duarte-Mermoud and Travieso-Torres (2003). This controller has the following form

$$u(t) = \theta_1(t)y + \theta_3(t)\varpi(t)$$

$$\dot{\theta}_1(t) = -\left(\Gamma_1^{-1}/\sqrt{1 + Trace\left(\Gamma_1^{-2} + \Gamma_3^{-2}\right)}\right)yy^T, \qquad \dot{\Gamma}_1 = -\Gamma_1 yy^T \Gamma_1, \qquad \Gamma_1(t_0) > 0 \quad (17)$$

$$\dot{\theta}_3(t) = -\left(\Gamma_3^{-1}/\sqrt{1 + Trace\left(\Gamma_1^{-2} + \Gamma_3^{-2}\right)}\right)y\varpi^T(t), \qquad \dot{\Gamma}_3 = -\Gamma_3 \varpi(t)\varpi^T(t)\Gamma_3, \qquad \Gamma_3(t_0) > 0$$

This controller will convert system (6) to an equivalent C^2-passive system from the input $\varpi(t)$ to the output $y(t)$. The parameters $\theta_1(t) \in \mathfrak{R}^{2\times8}$ and $\theta_3(t) \in \mathfrak{R}^{2\times2}$ represent adjustable controller parameters whose ideal values are $\theta_1 = -\Lambda_b^{-1}\Lambda_a \in \mathfrak{R}^{2\times8}$ and $\theta_3 = \Lambda_b^{-1} \in \mathfrak{R}^{2\times?}$.

2.5. Simulation results of APBC for the IM

In order to verify the advantages of the proposed controllers a comparison with a traditional current regulated PWM induction motor drive from Chee-Mun (1998) with PI loop controllers (see Figure 2), was carried out. In the simulations a squirrel-cage induction motor whose nominal parameters are: 15 [kW] (20 [HP], 220 [V], fp= 0.853, 4 poles, 60 [Hz], R_s = 0.1062 [Ω], $X_{ls}=X_{lr}$ = 0.2145 [Ω], x_m = 5.8339 [Ω], R_r = 0.0764 [Ω], J = 2.8 [kg m²] and B_p = 0 were considered (Chee-Mun, 1998). All the simulations were made using the software package SIMULINK/MATLAB with ODE 15s (stiff/NDF) integration method and a variable step size.

The obtained control schemes only need the exact values or the estimates of parameters X_m and T_r for the field orientation block. No other parameters or state estimations are used. The PI speed controller is tuned as P=30 and I= 10 according to Chee-Mun (1998).

Figure 3 shows the information used to compare both control schemes. The variations of the reference speed ω_r^* (Figure 3(a)), the variations in load torque (Figure 3(b)), the variation of about 30% in the stator and rotor resistance (Figure 3(c) and Figure 3(d)), the linear increase up to double the load inertia during the motor operation (Figure 3(e)) and the variations in the viscous friction coefficient (Figure 3(f)). For both proposed control strategies (CFAG and

CTVAG) and the classical FOC control (BCS), five comparative tests considering the variations shown in Figure 3 were carried out.

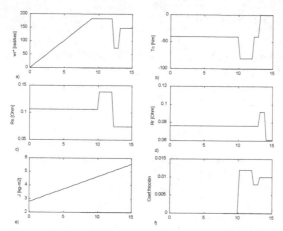

Figure 3. Parameter and reference variations used in the set of comparative tests

These tests allow us to study the behavior of the schemes under the situations described next.

- *Test 1:* The reference of speed is increased as a ramp from 0 to 190 [RPM] in 0.5 [s] and the load torque is fixed at the nominal value 69.5 [Nm].
- *Test 2:* Variations on load torque, as indicated in Figure 3(b).
- *Test 3:* Variations on speed reference, as shown in Figure 3(a).
- *Test 4:* Variation of the motor resistances, as shown in Figures 3(c) and 3(d).
- *Test 5:* Variation of the load parameters, as indicated in Figures 2(e) and 2(f).
- *Test 6:* Changes in the controller parameters (P and I) of the control loops.

In all the simulation results of the proposed controllers shown in Figure 4 through 9, the initial conditions of all the controller parameters and adaptive gains were set equal to zero, that is to say, $\theta_{ik}(0) = \gamma_{ih}(0) = 0$, for $i=1,2$ y $h=1,4$.

Figure 4 shows the comparative results obtained for the proposed controllers under normal conditions (i.e. according to Test 1), without considering variations of any type. APBC controllers present better transient behavior than traditional PI controllers. CFAG presents a quite accurate stationary state (with a velocity error less than 0.5 %). And CTVAG is equally accurate as the CFAG, but with better transient behavior.

Let us observe next in Figure 5, how the different schemes behave under variations of the load torque, as described in Figure 3(b). In the case of the CFAG shown, the error values are 0.5 % for a nominal load torque and of 0.22% for a half nominal load torque. The **CTVAG** presents a similar response to that of CFAG, but the transient response is slightly better. APBC controllers have better transient behavior than BCS.

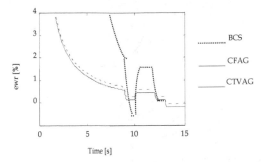

Figure 4. Results for the initial situation

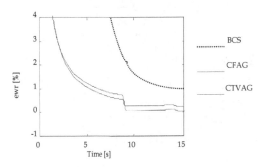

Figure 5. Results under load torque variations

In Figure 6, the effects of speed reference variations at nominal load torque, according to the variations indicated in Figure 3, are presented. The results for the proposed CFAG and CTVAG are similar rendering similar velocity errors whereas the rest of the variables present a suitable behavior. In these cases we have an error of about 0.5 % for nominal speed and of approximately 1.1 % at half the nominal speed.

When analyzing Test 4 (Figures 3(c) and 3(d)) both controllers present good behavior under changes on the stator resistance (see Figure 7). Nevertheless, under changes of the rotor resistance the field orientation is lost and the speed response is affected considerably. Notice how the flow of the machine diminishes considerably when the rotor resistance is decreased. We can also claim that the response in both cases (CFAG and CTVAG)is much more robust than the traditional PI controller of BCS. Both controllers present lesser speed errors in steady state than the classical PI scheme.

Considering now the variations of the load parameters according to Test 5 (Figures 3(e) and 3(f)), neither of the two controllers under study were affected, as is shown in Figure 8. For the proposed controllers, the differences found in the general behavior still remain. CFAG presents a similar error in the steady state than the CTVAG, but with slightly better transient behavior.

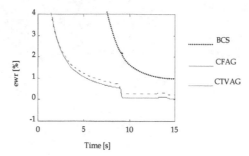

Figure 6. Results for speed reference variations

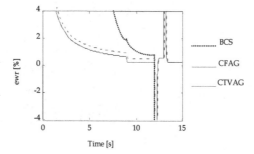

Figure 7. Results for Test 4.

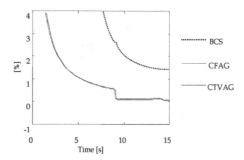

Figure 8. Results under for variations of load parameters

In Figure 9, the proportional gains of all control loops were changed. For CFAG and CTVAG, variations for the speed loop control parameter of 37.5 % were applied (P varies from 80 to 50). The flux loop was varied by 13 %, (P changes from 69 to 60). The current loops were varied by 33.3 % (P varies from 30 to 20). In Figure 9 it can be seen how in spite of these simultaneous gain variations, the speed error continues being less than 1% and the transient response after 0.5 sec. was practically not affected. CFAG as well as CTVAG guarantees good results for a wide range of variations of the proportional gains.

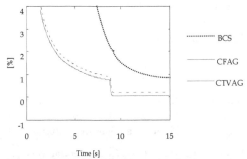

Time [s]

Figure 9. Results under changes in the tuning of the proportional gains

3. Control of IM using IDA-PCB techniques

In this section we will present a brief summary of the *Interconnection and Damping Assignment – Passivity-Based Control (IDA-PBC)* technique and the main ideas on which this method is based. This method provides a novel technique for computing the control necessary for modifying the storage function of a dynamical system assigning a new internal topology (in terms of interconnections and energy dissipation). Further details on the method can be found in Ortega et al. (2002) and Ortega & García-Canseco (2004). Next we will apply this technique to the control of an IM and compare it with BCS and APBC already described in Section 2.

3.1. Foundations of IDA-PCB control

Let us consider a system described in the form called Port-Controlled Hamiltonian (PCH) (Van der Shaft, 2000)

$$\Sigma_{PCH}: \begin{cases} \dot{x} = \left[J(x) - R(x)\right]\nabla H + g(x)u \\ y = g^T(x)\nabla H \end{cases} \qquad (18)$$

where $x \in \Re^n$ is the state, and $u, y \in \Re$ are the input and the output of the system. H represents the system's total stored energy, $J(x)$ is a skew-symmetric matrix ($J(x) = -J^T(x)$) called *Interconnection Matrix* and $R(x)$ is a symmetric positive definite matrix ($R(x) = R(x)^T \geq 0$) called *Damping Matrix*. Let us assume (Ortega et al., 2002; Ortega & García-Canseco, 2004) that there exist matrices $g^{\perp}(x)$, $J_d(x) = -J_d^T(x)$, $R_d(x) = R_d^T(x) \geq 0$ and a function $H_d : \Re^n \to \Re$, such that

$$g^{\perp}(x)\left[J(x) - R(x)\right]\nabla H = g^{\perp}(x)\left[J_d(x) - R_d(x)\right]\nabla H_d \qquad (19)$$

where $g^{\perp}(x)$ is the full-rank left annihilator of $g(x)$ ($g^{\perp}(x)g(x) = 0$) and $H_d(x)$ is such that $x^* = \arg\min_{x \in \Re^n}(H_d)$. Then, applying the control $\beta(x)$ defined as

$$\beta(x) = \left[g^T(x)g(x) \right]^{-1} g^T \left\{ \left[J_d(x) - R_d(x) \right] \nabla H_d - \left[J(x) - R(x) \right] \nabla H \right\} \tag{20}$$

the overall system under control can be written as

$$\dot{x} = \left[J_d(x) - R_d(x) \right] \nabla H_d \tag{21}$$

where x^* is a locally Lyapunov stable equilibrium. That is to say applying control (20) to (18) the dynamic of the system is changed to that shown in (21). x^* is a locally Lyapunov asymptotically stable equilibrium if it is an isolated minimum of H_d and the largest invariant inside the set $\left\{ x \in R^n \middle| \nabla H_d^T(x) R_d(x) \nabla H_d(x) \right\}$ is equal to $\{x^*\}$.

There are two ways to find control (20). The first one consists of fixing the topology of the system (by fixing J_d, R_d and $g^\perp$) and solving the differential equation (19). The second method consists of fixing H_d (the initial geometrical form of the desired energy) and then (19) becomes an algebraic system that has to be solved for J_d, R_d and $g^\perp$ (Ortega et al., 2002; Ortega & García-Canseco, 2004).

For the IDA-PBC scheme developed in Section 3.2, the model of the IM should be expressed in the PCH form previously stated, which has the general form shown in (18). In this study the load torque will be assumed proportional to rotor speed ($T_c = B\omega_r$) which typically represents fan load type. In this particular case the PCH model of the induction motor (see (22)), assuming also that the speed of the x-y reference system is synchronized to electrical frequency ($\omega_g = \omega_s$), has the form (González, 2005, González& Duarte-Mermoud, 2005; González et al., 2008)

$$\dot{x} = \begin{bmatrix} -R_s & 0 & 0 & 0 & 0 \\ 0 & -R_r & 0 & 0 & -x_4 \\ 0 & 0 & -R_s & 0 & 0 \\ 0 & 0 & 0 & -R_r & x_2 \\ 0 & x_4 & 0 & -x_2 & -B' \end{bmatrix} \nabla H + \begin{bmatrix} 1 & 0 & x_3 \\ 0 & 0 & x_4 \\ 0 & 1 & -x_1 \\ 0 & 0 & -x_2 \\ 0 & 0 & 0 \end{bmatrix} \begin{pmatrix} u_{sx} \\ u_{sy} \\ \omega_s \end{pmatrix},$$

$$y = \begin{bmatrix} 1 & 0 & 0 & 0 & 0 \\ 0 & 0 & 1 & 0 & 0 \\ x_3 & x_4 & -x_1 & -x_2 & 0 \end{bmatrix} \nabla H = \begin{bmatrix} i_{sx} \\ i_{sy} \\ 0 \end{bmatrix}$$

$$x = \begin{bmatrix} \psi_{sx} & \psi_{rx} & \psi_{sy} & \psi_{ry} & J\omega_r \end{bmatrix}^T = \begin{bmatrix} x_{12}^T & x_{34}^T & x_5 \end{bmatrix}^T,$$

$$H = \frac{1}{2} x_{12}^T L^{-1} x_{12} + \frac{1}{2} x_{34}^T L^{-1} x_{34} + \frac{1}{2} J^{-1} x_5^2$$

$$u = \begin{bmatrix} u_{sx} & u_{sy} & \omega_s \end{bmatrix}^T, \qquad y = \begin{bmatrix} i_{sx} & i_{sy} & 0 \end{bmatrix}^T,$$

$$\text{with} \qquad L = \begin{bmatrix} L_s & L_m \\ L_m & L_r \end{bmatrix} \tag{22}$$

where $\psi_{sx}, \psi_{sy}, \psi_{rx}, \psi_{ry}$ are the stator and rotor fluxes, respectively, and $B' = B_p + B$. In general, when using PCH representation, the obtained state variables are not necessarily the best choice for analysis and additional measurement/estimation may be needed in the controller implementation. Other types of load torque may also be considered in this analysis (e.g. constant, proportional to squared speed, etc.), in which case a slightly different PCH model will be obtained.

3.2. IDA-PBC strategy applied to the IM

The IDA-PBC strategy (Ortega et al., 2002; Ortega & García-Canseco, 2004) consists basically of assigning a new storage function to the closed-loop system, changing the topology of the system, in terms of interconnections and energy transfers between states. In the case of IM (González, 2005, González& Duarte-Mermoud, 2005; González et al., 2008), the controller is defined by some feasible solution for k_1, k_2 and k_3 of the following algebraic equation

$$L^{-1}x_{12} + \left(\begin{array}{c} k_1 \\ \dfrac{x_4}{x_2^{\,2} + x_4^{\,2}} k_3 \end{array} \right) = 0, \quad L^{-1}x_{34} + \left(\begin{array}{c} k_2 \\ \dfrac{x_4}{x_2^{\,2} + x_4^{\,2}} k_3 \end{array} \right) = 0, \quad J^{-1}x_5 + k_3 = 0 \tag{23}$$

From the third equation in (23), it is observed that an equilibrium point $x_5^* = \omega_r^*$ exists for ω_r defined as $\omega_r^* = -k_3$. For the other parameters (k_1, k_2) the solutions are given by the following relationship $(k_1^2 + k_2^2)L_m^{\,2} \geq 2k_3L_rB$ (González, 2005; González et al., 2008).
With the previous results, according to (20), the IDA-PBC controller is defined as

$$u_{sx}(x) = -R_s k_1 + \left(1 + \frac{R_r B'}{x_2^{\,2} + x_4^{\,2}} \right) x_3 k_3,$$

$$u_{sy}(x) = -R_s k_2 - \left(1 + \frac{R_r B'}{x_2^{\,2} + x_4^{\,2}} \right) x_3 k_3, \tag{24}$$

$$\omega_s(x) = -\left(1 + \frac{R_r B'}{x_2^{\,2} + x_4^{\,2}} \right) k_3$$

States x_2 and x_4 correspond to rotor flux expressed in orthogonal coordinates (ψ_{rx}, ψ_{ry}). The rotor flux will be zero if and only if the motor is at rest and without voltage applied. At t=0, some tension has to be applied to control the motor and therefore ψ_r becomes different from zero at t=0. Thus, no undetermined values of the controller are obtained.

The IDA-PBC scheme used in this paper was slightly modified. In principle, this strategy was developed to control the motor speed, not being robust with respect to load perturbations on the motor axis. This means that permanent errors in the mechanical speed were obtained. In order to solve this problem, a simple proportional integral loop was added for the speed error loop modifying the original IDA-PBC, scheme as is shown in Figure 10.

In general the rotor flux cannot be measured in the majority of IM's, which is why it was necessary to implement a rotor flux observer for the experimental implementation of this strategy. The observer was implemented based on the voltage-current model of the induction motor, developed in Marino et al (1994), Jansen et al (1995) and Martin (2005).

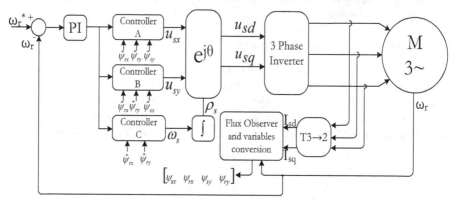

Figure 10. The IDA-PBC control scheme

3.3. Simulation results using IDA-PCB

In this section we present simulation results of applying the IDA-PBC technique for the speed control of an IM (Pelisssier, 2006; Pelisssier & Duarte-Mermoud, 2007). These results are compared with the basic control strategy (BCS) described in Figure 2 and with APBC strategies with fixed and time varying adaptive gains described in Section 2 (Figure 1). The results were obtained using Matlab/Simulink and the IM considered is that described in Chee-Mun (1998). The following two tests were performed on the simulated IM.

Test 1 (Regulation): Speed ramp from zero to nominal speed in 20 seconds with load torque proportional to speed, staring from zero. Then in t=25[s] a load torque of 50% magnitude of the nominal torque value is applied; in t=40[s] the magnitude of the load torque is increased to 100%; in t=60[s] the magnitude of the load torque is decreased to 50% and finally in t=120[s] the load torque is set to zero.

Test 2 (Tracking): Speed ramp from zero to nominal speed in 20 seconds with load torque proportional to speed, staring from zero. Between t=50[s] and 100[s] a pulse train of amplitude $0.1\omega_{r_{nom}}$ and frequency $2\pi/20$ is added to the constant speed reference. Between t=120[s] and 160[s] a sinusoidal speed reference of amplitude $0.1\omega_{r_{nom}}$ and frequency $2\pi/20$ is added to the constant nominal speed reference. The load torque is kept constant in 50% of the nominal torque during the whole test.

The PI controller parameters were first determined using the Ziegler-Nichols criteria and modified later by simulations, until a good response was obtained. For the APB scheme the controllers 'constants were chosen as follows: K_P=0.3 and K_I=0.1 for the external loop and

$K_P=500*76.82$ for the internal loop. For the IDA-PBC scheme, the values of the parameters were chosen so that equation (23) is satisfied. The values found were $k_1=k_2=-7$. For the external loop the values were chosen as $K_P=K_I=0.5$. The results were compared with the BCS described in Figure 2 and the APBC shown in Figure 1.

The simulations results obtained for Test 1 and Test 2 are shown in Figures 11 and 12.

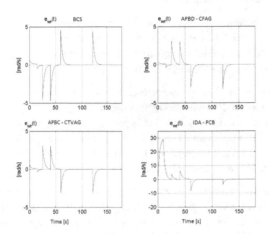

Figure 11. Simulation results for Test 1

The results obtained from Test 1 (Figure 11) show that the smaller errors are obtained by APBC strategies (CFAG and CTVAG) with a maximum error around 3 [rad/s]. This error is less than those obtained from the BCS and the IDA-PBC strategies which are around 5 and 30 [rad/s] respectively. However, the settling time of all four strategies is similar.

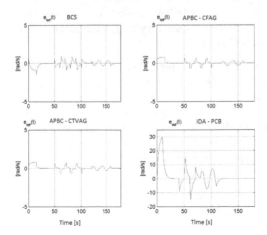

Figure 12. Simulation results for Test 2

From the results obtained for Test 2 (Figure12) a faster stabilization is obtained by the APBC strategies (CFAG and CTVAG), followed by the BCS strategy which was better than the IDA-PBC. The later is strongly dependant on the dynamics of the external loop introduced for controlling the mechanical torque.

3.4. Experimental results using IDA-PBC

In this section the experimental results obtained by applying APBC (CFAG and CTVAG) described in Figure 2 and IDA-PBC strategies described in Figure 10, are presented and compared with the BCS described in Figure 3. The experimental set up as well as the tests carried out for each strategy are described in what follows.

The three phase inverter used in the experiments was that designed and built by González (2005). Communication to PC was done though the software Matlab-Simulink using a customized S-Function. The IM used in the experiments was a Siemens 1LA7080, 0.55KW, $\cos(\phi)$=0.82, 220V, 2.5A, 4 poles and 1395RPM. From motor tests (no load and locked rotor) the estimated motor parameters used in the study the following: Rs=14.7Ω, Rr=5.5184Ω, Xs=11.5655Ω, Xr=11.5655Ω and Xm=115.3113Ω.

In order to apply resistive torque on motor axis, the induction motor was mechanically coupled to a continuous current generator, Briggs & Stratton ETEK, having a permanent magnet field. The load to the generator was applied using a cage of discrete resistances connected to generator stator and manually controlled by switches. The magnitudes of the resistances were chosen such that maximum values of induction motor operation were not exceeded under any circumstances. The experimental assembly including the motor-generator group used in the experimental tests is shown in Figures 13 (a) and (b).

Test 1 (Basic Behavior): The speed reference was a ramp starting from zero at t=0 to the nominal speed (146.08 rad/s) in 9s. The load torque was kept constant and equal to the nominal value (100%) during the whole test. Initial conditions (IC) for controller parameters were all set to zero, except for the time-varying gains which were chosen as $\Gamma_1 (0)= \Gamma_3 (0)=I$, where I is the 2x2 identity matrix.

Test 2 (Tracking): A ramp speed referenced was considered, starting from rest at zero and reaching the nominal speed (146.08 [rad/s]) in 9[s]. Between t=40[s] and t=70[s] a pulse train reference of amplitude $0.1\omega_{r_{nom}}$ and frequency $\pi/10$ [rad/s] was added on top of the constant nominal value. Between t=80[s] and t=110[s] a sinusoidal reference of amplitude $0.1\omega_{r_{nom}}$ and frequency $\pi/10$ was added on top of the constant nominal value. Additionally, the load torque (proportional to the speed) was kept at 50% of the nominal during the whole test. The IC of the controller parameters were all set to zero, except for time-varying gains initial values that were chosen as $\Gamma_1 (0)= \Gamma_3 (0)=I$, where I is the 2x2 identity matrix.

Test 3 (Regulation): The speed reference was a ramp starting from rest at zero reaching the nominal value(146.08 [rad/s]) in 9s, where the reference was kept constant. Initial load torque was equals to 0% of the nominal value. Between t=40[s] and t=80[s] a torque perturbation equal to 50% of the nominal value is added.

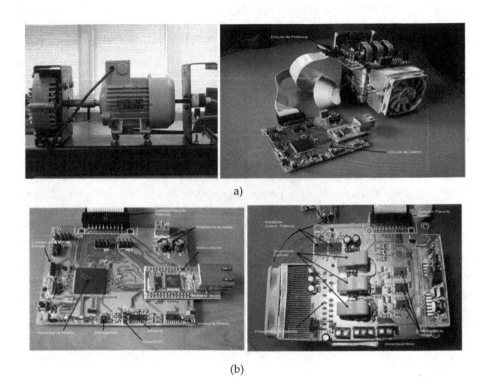

a)

(b)

Figure 13. (a). Experimental assembly. Motor-generator and inverter. (b). Experimental assembly. Control circuit and power circuit.

For the experimental tests, the best values of PI controller parameters for inner and outer loops were chosen based on those obtained from the simulation results of Section 3.2 (González, 2005; González& Duarte-Mermoud, 2005; Pelissier & Duarte-Mermoud, 2007). Later, these values were adjusted during the experiments performing a small number of trial tests. The final values chosen for the constants of control loops used in the BCS and in APBC scheme are as follows: K_P=0.403 and K_I=0.0189 for the outer loop and K_P=45 for the inner loop. For the IDA-PBC strategy, the values of constants k_1 and k_2 were determined based on simulations results reported in Pelissier & Duarte-Mermoud (2007). The chosen values were k_1=k_2=-30 and for the proportional integral loop it K_P=3 and K_I=0.5 were chosen.

For the experimental tests the control strategies were implemented in Matlab/Simulink, using a fixed step of 10[micro s] and the solver ODE5 (Dormand-Prince). In the electronics, a vector modulation with a carrier frequency of 20[kHz] was used. All IC were set to zero except time-varying gains initial values which were chosen as Γ_1 (0)= Γ_3 (0)=I, where I is the 2x2 identity matrix.

The experimental results obtained after applying the techniques under study for Test 1, Test 2 and Test 3 already described, are shown next. In Figures 14 through 16 the evolution of the speed errors are plotted for each strategy for each one of the tests.

In Figure 14 it is observed the following results: the fastest convergence of control error to zero, with a constant nominal load torque applied (Test 1), was achieved by the IDA-PBC strategy, with about 40[s]. Then 60[s] and 80[s] were obtained by BCS and APBC strategies, respectively. However, in the IDA-PBC strategy an important oscillatory behavior of the control error is observed at the beginning. For more information about the behavior of other variables see González (2005), Pelissier & Duarte-Mermoud (2007) and González et al (2008).

From the tracking viewpoint (Test 2), the best results were achieved for the APBC strategies, which follow reference changes better than the BCS (See Figure 15). The IDA-PBC strategy is not able to follow reference changes properly, presenting an oscillatory behavior of speed error. Convergence of the control error to zero for the IDA-PBC is influenced by rotor flux observer convergence, which necessarily adds a dynamic to the system affecting the global behavior of the overall system. For information about the evolution of other variables see González (2005), Pelissier & Duarte-Mermoud (2007) and González et al (2008).

When applying torque perturbations on the motor axis (Test 3), it is observed that fastest stabilization was attained by APBC strategies, without large oscillations (see Figure 16). The IDA-PBC strategy, although perturbations are quickly controlled, has an oscillatory control error. The BCS case is the slowest with a larger error in stationary state. This last strategy is not robust in the presence of perturbations on the mechanical subsystem. The evolution of other variables can be seen in (González, 2005; Pelissier & Duarte-Mermoud, 2007; González et al., 2008).

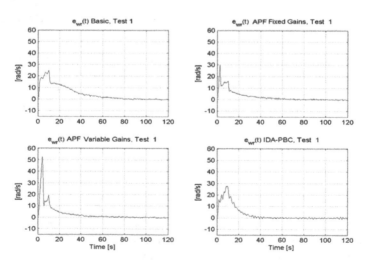

Figure 14. Speed errors for experimental Test 1 with constant load torque

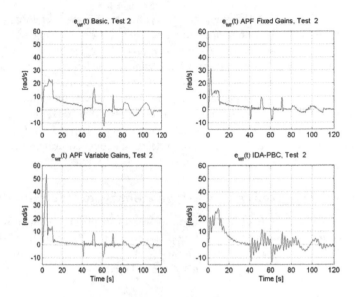

Figure 15. Speed errors for experimental Test 2 for reference tracking

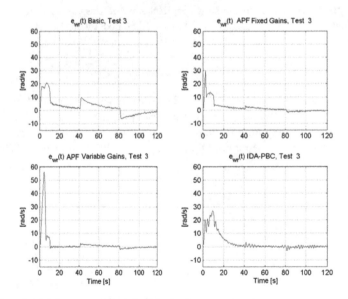

Figure 16. Speed errors for experimental Test 3 for load torque perturbations

Numerous other experiments and simulations, not shown here for the sake of space, were carried out to analyze the influence of several other parameters on the BCS, APBC and IDA-

PBC strategies (Travieso, 2002; Pelisssier, 2006). In particular the effects of initial conditions on APBC strategies, as well as the effects of using fixed and time-varying adaptive gains were analyzed. It was observed, in general, that time-varying gains improve transient behavior and diminish initial control error. In simulations, a small noise was added on these signals and the performance of the method were not affected significantly (González, 2005). At the experimental level, the influence of the normal noise present in the measurement of current signals during the test did not affect the behavior of the APBC. For higher noise levels some deterioration of the control system behavior was observed. In this case, more robust adaptive laws should be used. For instance the σθ-modification (Narendra & Annaswamy, 1989) could be used.

4. Induction motor speed control using fractional order PI controllers

In this section we present a field oriented control scheme like the one shown in Figure 2, where the PI controller used in the speed loop is changed to a PI controller in which the integral order is not unity (fractional integral effect). The main idea to explore is that fractional order integrals are of benefit in this kind of IM controllers.

4.1. Fractional order PI controllers

The FOPI controller is based on the same principles as the classical PI controller, with the difference that in this case the control action is calculated by means of fractional order integrals. The transfer function of a FOPI controller is given by

$$H_{FOPI}(s) = k_p + k_i / s^v \tag{25}$$

where v denotes the integration order, k_p is the proportional constant and k_i is the integral constant. The detailed computation of fractional integrals is shown (Valério, 2005). Expression (8) allows computing fractional integrals $v < 0$ and fractional derivatives $v > 0$, corresponding to Caputo's definition (Oldham & Spanier, 1974; Kilbas et al., 2006; Sabatier et al., 2007).

$$\frac{d^v}{dx^v} = \begin{cases} \dfrac{1}{\Gamma(-v)} \displaystyle\int_{x_0}^{x} (x-t)^{-v-1} f(t)dt, & v < 0 \\[3mm] f(x), & v = 0 \\[3mm] \dfrac{1}{\Gamma(-v)} \displaystyle\int_{x_0}^{x} (x-t)^{-v-1} \left[D^m f(t) \right] dt, & v > 0 \end{cases} \tag{26}$$

In this equation x corresponds to the integration variable, Γ corresponds to Gamma function, m denotes the integer immediately greater than v, and D denotes the integer derivative with respect to x. The Laplace Transform of fractional order derivatives and integrals (according to Caputo's definition) is shown in (9), where $F(s) = \mathcal{L}\{f(x)\}$.

$$L\left\{\frac{d^{v}}{dx^{v}}f(x)\right\} = \begin{cases} s^{v}F(s), & v \leq 0 \\ s^{v}F(s) - \sum_{k=0}^{n-1} s^{k}\left({}_{0}^{x}D^{v-k-1}\right)f(0), & v > 0 \end{cases} \tag{27}$$

Other definitions commonly used in fractional calculus can be found in Oldham & Spanier (1974), Valério (2005), Kilbas et al (2006) and Sabatier et al (2007).

In this study the FOC scheme shown in Figure 2 (changing the PI controllers by FOPI controllers in the spped control block) is used to control the speed of an IM. The "Speed Controller" block shown in Figure 2 corresponds to a FOPI controller, in which the parameter v will be modified to analyze its effects on the controlled system. This strategy will be denoted as FOC-FOPI and will be compared with the classical strategy using a standard PI controller (BCS), which will be denoted as FOC-PI. Notice that this case corresponds to the FOC-FOPI strategy when $v = 1$. "Current Controller" block and "Flux Controller" block in Figure 2 correspond to proportional controllers. All these controller parameters (proportional constants) will be kept constant at values indicated in Section 4.2.

4.2. Simulation set up

The controlled system corresponds to the Siemens 3-phase IM, model 1LA7080 descrbed in Section 3.3. All the simulations shown in this study were performed using MATLAB/Simulink. The following describes the tests performed on the IM, in simulation analysis. These tests were used to determine the general features of FOC-FOPI scheme, and will be compared with FOC-PI scheme. For the sake of space only results concerning the regulation of the controlled system (capacity of the system reject external perturbations at different levels of load) are shown. Although the tracking study (capacity of the system to reach and follow a pre-specified speed reference at different levels of load) was also done (Mira, 2008; Mira & Duarte-Mermoud, 2009; Duarte-Mermoud et al, 2009) the simulated results are not shown here.The simulation results will be analyzed and discussed including stabilization time, rise time and control effort, among other aspects.

Test 1(Regulation): The speed reference increases from zero at a rate of 16 $[rad/s^{2}])$ until nominal speed ($146.08[rad/s] = 1395[rpm]$) in $9.33[s]$. Then the reference is kept constant at the nominal speed until the end of the test ($315[s]$). (See Figure 4). The mechanical load varies during the test as shown in Table 2 and Figure 17.

The FOPI controller has the transfer function shown in (7). In all tests, parameters k_p and k_i were kept fixed at 0.5 and 0.05 respectively. These values were chosen after performing a series of preliminary simulation tests, analyzing the stabilization time and the control effort for different values. The values of the proportional constant used in current and flux proportional controllers were chosen to be 45. This value was also determined after a series of preliminary tests. The integration order was changed to explore the system's sensitivity with respect to parameter v. The results shown in the next section include orders 0.7, 1.0 (Classical PI), 1.7 and 2.0. Theoretically the limit of stability is at $v = 2$, a fact verified at the simulation level (Mira & Duarte-Mermoud, 2009). See also Figure 21. Many other integration orders have been analyzed at the simulation level but they are not shown here for the sake of space. The reader is referred to Mira (2008) for more details.

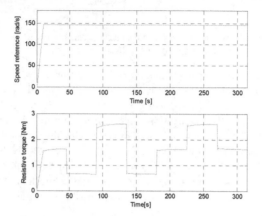

Figure 17. Description of Test 1 (Regulation)

Interval [s]	Level of load
0 – 45	50%
45 – 90	0%
90 – 135	100%
135 – 180	0%
180 – 225	50%
225 – 270	100%
270 – 315	50%

Table 1. Variation of load torque

4.3. Simulation results (regulation)

In this section the simulated behavior of IM control under the FOC-FOPI scheme, when Test 1 is applied, is shown in Figs. 18 to 21, for different values of the integration order ν of the integral part of PI controller (ν = 0.7, ν = 1.0, ν = 1.7 and ν = 2.0). Note that the classical FOC-PI scheme is obtained from the FOC-FOPI strategy by setting ν = 1. In all the figures, only the controlled variable (motor speed) is shown. The results obtained are quite satisfactory, as can be seen from Figures 18 to 21. The evolution of the controlled variable tries to follow the speed reference at all times, in spite of the perturbation being applied.

From Figure 18 it can be observed that for integration orders less than 1.0 the response presents no overshoot although the response is slower. It can be concluded, from information contained in Figure 20, that the response is faster when the integration order is greater than 1.0 but an overshoot is observed. When ν is chosen as 2.0 critically stable behavior is attained, as shown in Figure 21

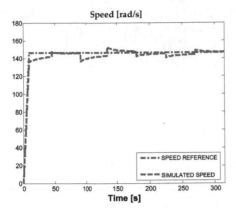

Figure 18. Simulation results for $v = 0.70$

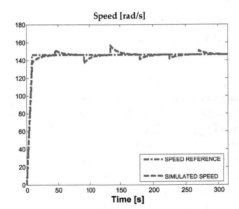

Figure 19. Simulation results for $v = 1.00$

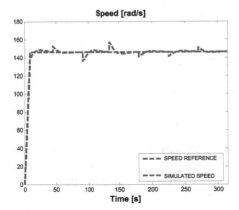

Figure 20. Simulation results for $v = 2.00$

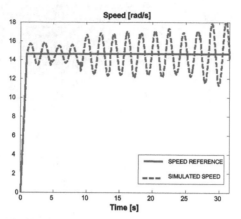

Figure 21. Simulation results for $v = 1.30$

Experimental analysis of the FOC-FOPI scheme is currently underway. These results will be reported, compared and discussed in the near future.

5. Conclusions

From simulation and experimental analysis performed on induction motor control some interesting conclusions can be drawn. In APBC strategies an important simplification of control scheme based on FOC principle can be attained when using the TFCP, allowing an effective control of the system without the necessity of having a rotor flux sensor or implementing a rotor flux observer to orientate the field. For APBC strategy the use of time-varying adaptive gains noticeably improves the transient behavior of controlled system, both for tracking as well as for regulation, when compared with the APBC strategy with fixed adaptive gains and also when compared with classical control strategies. Results are very similar for both the SISO and the MIMO approaches when using APBC. Compared with other control schemes proposed in the literature such as those based on traditional PI (Chee-Mun, 19998), sliding modes (Chan & Wang, 1996 ; Dunningan et al, 1998 ; Taoutaou & Castro-Linares, 2000; Araujo & Freitas, 2000), artificial intelligence (Bose, 1997 ; Vas, 1999) and non adaptive passivity (Taoutaou & Castro-Linares, 2000; Espinosa & Ortega, 1995), we have been able to develop four simple and novel controllers. They have adaptive characteristics, being robust in the presence of load parameter variations. Simple proportional controllers are used for the rotor speed, rotor flux and stator current control loops. They are also robust for a large range of proportional gain variations.

In the case of energy shaping strategy, IDA-PBC, a novel control scheme was studied and implemented. Since the original strategy was only designed for speed control, the addition of an outer speed loop of proportional-integral type, allowed obtaining certain robustness with respect to torque perturbations. In this strategy was necessary the design and implementation of a rotor flux observer, adding certain complexity to the complete system.

Since the BCS has fixed controller parameters its behavior is not as good as the adaptive strategies studied and presented here.

From simulation results obtained in this study it is possible to state that the integration order of FOPI controller plays a central role in speed control of an IM, when compared with the BCS. Choosing a suitable value of the integration order allows obtaining fast/slow responses and over/under damped responses. For this particular case of IM speed control , it was observed that for integration orders lesser than 1 the stabilization time is rather large and the controlled variable may not present overshoot. On the contrary, for values over 1 the stabilization time is small (and diminishes as integration order increases); the overshoot increases as the integration order does, reaching instability for integration order equal to 2. It was observed that the best results obtained from this study correspond to integration orders near to 1.40, presenting small rise and stabilization times, though with certain degree of overshoot.

In conclusion, the adaptive strategies studied present clear advantages with respect to the BCS used as basis of comparison. Amongst the adaptive schemes the APBC with time-varying adaptive gains is the one that behaves better.

Author details

Manuel A. Duarte-Mermoud & Juan C. Travieso-Torres
Department of Electrical Engineering, University of Chile, Santiago, Chile

Acknowledgement

The results reported here have been supported by CONICYT-Chile under grants Fondecyt 1061170, Fondecyt 1090208 and FONDEF D05I-10098.

6. References

Al-Nimma, D.A. and Williams, S. (1980). Study of rapid speed-changing methods on A.C. motor drives. *IEE Proceedings, Part B, Electric Power Applications*, Vol.127, No.6, pp. 382 – 385, ISSN: 0143-7038

Araujo, R. E. & Freitas, D. (2000), "Non-linear control of an induction motor : sliding mode theory leads to robust and simple solution ". *International Journal of Adaptive Control and Signal Processing*, Vol.14, No. 2, pp. 331-353, MES, ISSN: 0890-6327.

Bose, B.K. (1997). *Power Electronics and Variable Frequency Drives; Technology and Applications*. IEEE Press Marketing, ISBN-13: 978-0780310841, New York, USA

Bose, B.K. (2002). *Modern Power Electronics and AC Drives*, Prentice Hall PTR, ISBN-13: 978-0130167439, Upper Saddle River, USA

Byrnes, C.I., Isidori, A. & Willems, J.C. (1991), "Passivity, feedback equivalence, and the global stabilization of minimum phase nonlinear systems". *IEEE Transactions on Automatic Control*, Vol. 36, No. 11, pp. 1228-1240, November, ISSN: 0018-9286.

Castro-Linares, R. & Duarte-Mermoud, M.A. (1998), "Passivity equivalence of a class of SISO nonlinear systems via adaptive feedback". *Proceedings of VIII Latinamerican Congress on Automatic Control*, 9 – 13 November, Vol. 1, pp. 249-254.

Chan, C.C. & Wang, H.Q. (1996), "New scheme of sliding-mode control for high performance induction motor drives". *IEE Proceedings on Electrical Power Applications*, Vol. 143, pp. 177-185, ISSN: 1350-2352.

Chee-Mun, O. (1998), *Dynamic Simulation of Electric Machinery, using Matlab/Simulink*. Prentice Hall PTR, ISBN-10: 0137237855, ISBN-13: 978-0137237852, USA.

Duarte-Mermoud M.A., Castro-Linares R. and Castillo-Facuse A. (2001), "Adaptive passivity of nonlinear systems using time-varying gains". *Dynamics and Control*, Vol. 11, No. 4, December, pp. 333-351, ISSN: 0925-4668.

Duarte-Mermoud, M.A. & Travieso-Torres, J.C. (2003), "Control of induction motors: An adaptive passivity MIMO perspective". *International Journal of Adaptive Control and Signal Processing*, Vol. 17, No. 4, May, pp. 313-332, ISSN: 0890-6327.

Duarte-Mermoud, M.A., Castro-Linares, R. & Castillo-Facuse, A. (2002), "Direct passivity of a class of MIMO nonlinear systems using adaptive feedback". *International Journal of Control*. Vol.75, No. 1, January, pp. 23-33, ISSN: 0020-7179.

Duarte-Mermoud, M.A., Méndez-Miquel, J.M., Castro-Linares, R. & Castillo-Facuse, A. (2003), "Adaptive passivation with time-varying gains of MIMO nonlinear systems". *Kybernetes*, Vol. 32, Nos. 9/10, pp. 1342-1368, ISSN: 0368-492X.

Duarte-Mermoud, M.A., Mira, F.J., Pelissier, I.S. & Travieso-Torres, J.C., "Evaluation of a fractional order PI controller applied to induction motor speed Control". *Proceedings of the 8th IEEE International Conference on Control & Automation (ICCA2010)*, June 9-11, 2010, Xiamen, China. Proc. in CD, Paper No. 790, ThA4.5, ISBN: 978-1-4244-5195-1.

Dunnigan, M.W., Wade, S., Williams, B.W. & Yu, X. (1998), "Position control of a vector controlled induction machine using Slotine's sliding mode control approach". *IEE Proceedings on Electrical Power Applications*, Vol. 145, pp. 231-248, ISSN: 1350-2352.

Espinosa, G. and Ortega, R. (1995), "An output feedback globally stable controller for induction motors". *IEEE Transaction on Automatic Control*, Vol. 40, pp. 138-143, ISSN: 0018-9286.

González, H. (2005). *Development of Control Schemes based on Energy Shaping for a Class of Nonlinear Systems and Design of an Open Three Phase Inverter for on-line Applications in Induction Motors*. (In Spanish). M.Sc. Thesis, Department of Electrical Engineering, University of Chile

González, H., Duarte-Mermoud, M.A., Pelissier, I., Travieso J.C. & Ortega, R. (2008). A novel induction motor control scheme using IDA-PBC. *Journal of Control Theory and Applications*. Vol. 6, No. 1, January 2008, pp. 123-132, ISSN: 1993-0623.

González, H.A. & Duarte-Mermoud, M.A. (2005a). Induction motor speed control using IDA-PCB". (In Spanish). *Anales del Instituto de Ingenieros de Chile (Annals of the Chilean Institute of Engineers)*, Vol.117, No.3, December 2005, pp. 81-90 ISSN 0716-3290.

Isidori, A. (1995), *Nonlinear Control Systems*. Springer Verlag, Third Ed, ISBN-10: 3540199160, ISBN-13: 978-3540199168, USA.

Jansen, P.L.; Lorenz, D. & Thompson, C.O. (1995). Observer-based direct field orientation for both zero and very high speed operation. *IEEE Ind. Applic. Society Magazine,* Vol.1, No.4, (Jul. 1995), pp. 7-13, ISSN 1077-2618.

Kilbas, A. A.; Srivastava, H. M. & Trujillo, J. J. (2006). *Theory and Applications of Fractional Differential Equations.* Elsevier Science Inc., North-Holland Mathematics Studies, Vol.204, ISBN 978-0-444-51832-3, New York, USA

Marino, R.; Peresada, S. & Tomei, P. (1994). Adaptive observers for induction motors with unknown rotor resistance. *Proceeding of the 33rd Conf. on Decision and Control,* Vol.1, pp. 4018-4023, ISBN 0-7803-1968-0, Orlando, Florida, USA, December 1994

Martin, C. (2005). *A Comparative Analysis on Magnetic Flux Observers for Induction Motor Control Schemes.* (In Spanish). E.E. Thesis, Department of Electrical Engineering, University of Chile

Mira, F. (2008). *Design of a Three Phase Inverter and its Application to Advanced Control of Electrical Machines.* (In Spanish), E. E. Thesis, Dept. Elect. Eng., University of Chile

Mira, F.J. & Duarte-Mermoud, M.A. (2009). "Speed control of an asynchronous motor using a field oriented control scheme together with a fractional order PI controller. (In Spanish). *Annals of the Chilean Institute of Engineers,* Vol.121, No.1, pp. 1-13, ISSN 0716-3290.

Narendra, K.S. & Annaswamy, A.S. (1989). *Stable Adaptive Systems.* Prentice Hall, ISBN 0-13-839994-8, Englewood Cliffs, New Jersey, USA

Nijmeijer, H. & Van der Schaft, A. (1990), *Nonlinear Dynamical Control Systems.* Springer Verlag, ISBN-10: 038797234X, ISBN-13: 978-0387972343, USA.

Oldham, K. B. & Spanier, J. (1974). *The Fractional Calculus: Theory and Applications of Differentiation and Integration to Arbitrary Order.* Academic Press, Inc., Mathematics in Science and Engineering, Vol.111, ISBN 0-12-558840-2, New York, USA.

Ortega, R. & García-Canseco, E. (2004). Interconnection and damping assignment passivity-based control: A survey. *European Journal of Control,* Vol.10, No.5, pp. 432-450, ISSN 0947-3580.

Ortega, R., van der Schaft, A., Maschke, B. & Escobar, G. (2002). Interconnection and damping assignment passivity-based control of port-controlled Hamiltonian systems. *Automatica,* Vol.38, pp. 585-596, ISSN 0005-1098.

Pelissier, I. & Duarte-Mermoud, M.A. (2007). Simulation comparison of induction motor schemes based on adaptive passivity and energy shaping. (In Spanish). *Anales del Instituto de Ingenieros de Chile (Annals of the Chilean Institute of Engineers),* Vol.119, No.2, August 2007, pp. 33-42, ISSN 0716-3290.

Pelissier, I. (2006). *Advanced strategies for induction motor control.* (In Spanish). E.E. Thesis, Department of Electrical Engineering, University of Chile

Sabatier, J.; Agrawal, O. P. & Machado, J. A. Eds. (2007). *Advances in Fractional Calculus: Theoretical Developments and Applications in Physics and Engineering.* Springer Verlag, ISBN ISBN 978-90-481-7513-0, New York, USA.

Taoutaou, D. & Castro-Linares, R. (2000), "A controller-observer scheme for induction motors based on passivity feedback equivalence and sliding modes". *International Journal of Adaptive Control and Signal Processing,* Vol. 14, No. 2-3, pp. 355-376, ISSN: 0890-6327.

Travieso, J.C. (2002). *Passive equivalence of induction motors for control purposes by means of adaptive feedback*. (In Spanish). Ph.D. Thesis, Electrical Engineering Department, Universidad de Santiago de Chile

Travieso-Torres, J.C. & Duarte-Mermoud, M.A. (2008), "Two simple and novel SISO controllers for induction motors based on adaptive passivity". *ISA Transactions*, Vol. 47, No. 1, January, pp.60-79, ISSN 0019-0578.

Valério, D. (2005). *Fractional Robust System Control*. Ph.D. Dissertation, Instituto Superior Técnico, Universidade Técnica de Lisboa, Portugal

Van der Schaft, A. (2000). *L2-Gain and Passivity Techniques in Nonlinear Control*. 2nd Edition. Springer-Verlag, ISBN: 1-85233-073-2, London, GB

Vas, P. (1998). *Sensorless Vector and Direct Torque Control*. Oxford University Press, ISBN-10: 0198564651, N-13: 978-0198564652, New York, USA

Vas, P. (1999). *Artificial-Intelligence-Based Electrical Machines and Drives*. Oxford University Press, ISBN-10: 019859397X, ISBN-13: 978-0198593973, USA.

Williams, B.W. & Green, T.C. (1991). Steady state control of an induction motor by estimation of stator flux magnitude. *IEE Proceedings, Part B, Electric Power Applications*, Vol.138, No.2, pp. 69 -74, ISSN: 0143-7038

Industrial Application of a Second Order Sliding Mode Observer for Speed and Flux Estimation in Sensorless Induction Motor

Sebastien Solvar, Malek Ghanes, Leonardo Amet,
Jean-Pierre Barbot and Gaëtan Santomenna

Additional information is available at the end of the chapter

1. Introduction

Recently, considerable research efforts are focused on the sensorless Induction Motors (*IM*) control problem. We refer the reader to [12] for a tutorial account on the topic. Indeed, industries concerned by sensorless *IM* drives are continuously seeking for cost reductions in their products. The main drawback of *IM* is the mechanical sensor. The use of such direct speed sensor induces additional electronics, extra wiring, extra space, frequent maintenance, careful mounting and default probability. Moreover, the sensor is vulnerable for electromagnetic noise in hostile environments and has a limited temperature range.

To avoid mechanical sensor (speed, position and load torque) of *IM*, several approaches for the so-called "sensorless control" have attracted a great deal of attention recently (see for example [21], [15], [22], [16], [11], [14], [6], [10], [1], [8], [19]. These methods can be classified into three main strategies.

- Artificial intelligence strategies [22], [19].
- Strategies based on IM spatial saliency methods with fundamental excitation and high frequency signal injection [16], [12].
- Fundamental motor model strategies: adaptive observer [21], Luenberger observe [15], Kalman filter observer [11], high gain observer [14], [6], sliding mode observer [10], [1], interconnected high gain observer [8].

This chapter belongs to the third strategy using mainly observer methods.

First and second strategies have been a subject of growing interest in recent years. For example the second strategy based on IM spatial saliency with extra converters is a robust and physical method. But artificial intelligence and spatial saliency algorithms are quite heavy for basic microprocessors.

The third strategy that is a powerful observer that can estimate simultaneously variables and parameters of a large class of nonlinear systems doesn't require a very high performance processor for real time implementation but they are often tested at high speed in sensorless *IM* whereas the main difficulties are mainly at very low frequencies [10], [8].

However for our best of knowledge, examination of the literature on the third strategy shows that the real time computation constraints with a cheapest microprocessors or microprocessors not specially allowed to this task[1] are not taken into account to deal with industrial applications of sensorless *IM* including very low frequencies drives.

Meanwhile, compared with other observers, sliding mode technic [20] have attractive advantages of robustness against matching disturbances and, insensitivity to some specific variation of parameters in sliding mode behavior. However, the chattering effect (that is inherent to standard first sliding mode technic) is often an obstacle for practical applications. Higher-Order Sliding Modes (see for example [2], [18] and [5]) are one of the solutions which does not compromise robustness and avoid filtering of estimated variables as considered by other methods.

In this chapter, a second order sliding mode observer for the *IM* without mechanical sensor is presented for the open problem of sensorless *IM* drives at very low frequency. This observer converges in finite time and is robust to the variation of parameters. To illustrate the proposed observer, firstly a very simple case is presented in order to exemplified the tuning parameters. Then, to highlight the technological interest of the proposed method and also show the difficulties due to real time computation constraints when a basic microprocessors are used, an industrial application is proposed.

This paper is organized as follows: the section 2 recalls both *IM* model and unobservability phenomena of *IM*. In section 3 the super twisting algorithm (second sliding mode observer) is first presented in a simple case and then applied for sensorless *IM*. After that the section 4 proposes a discrete version of the super twisting observer. In section 5 the experimental results of the proposed observer carried out in an industrial framework are presented. Some conclusions and remarks are drawn in section VII.

2. Technical background

2.1. *IM* model

In [4] the following *IM* model is proposed, in the fixed (α, β) frame:

$$
\begin{cases}
\dot{i}_{s\alpha} = -\dfrac{R_s L_r^2 + R_r M_{sr}^2}{\sigma L_s L_r^2} i_{s\alpha} + \dfrac{M_{sr}}{\sigma L_s L_r} \left(\dfrac{R_r}{L_r} \phi_{r\alpha} + p\,\Omega\phi_{r\beta} \right) + \dfrac{1}{\sigma L_s} v_{s\alpha} \\[2ex]
\dot{i}_{s\beta} = -\dfrac{R_s L_r^2 + R_r M_{sr}^2}{\sigma L_s L_r^2} i_{s\beta} + \dfrac{M_{sr}}{\sigma L_s L_r} \left(\dfrac{R_r}{L_r} \phi_{r\beta} - p\,\Omega\phi_{r\alpha} \right) + \dfrac{1}{\sigma L_s} v_{s\beta} \\[2ex]
\dot{\phi}_{r\alpha} = \dfrac{M_s r R_r}{L_r} i_{s\alpha} - \dfrac{R_r}{L_r} \phi_{r\alpha} - p\,\Omega\phi_{r\beta} \\[2ex]
\dot{\phi}_{r\beta} = \dfrac{M_s r R_r}{L_r} i_{s\beta} - \dfrac{R_r}{L_r} \phi_{r\beta} + p\,\Omega\phi_{r\alpha} \\[2ex]
\dot{\Omega} = \dfrac{p M_s r}{J L_r} \left(\phi_{r\alpha} i_{s\beta} - \phi_{r\beta} i_{s\alpha} \right) - \dfrac{f}{J} \Omega - \dfrac{1}{J} T_l
\end{cases}
\tag{1}
$$

---

[1] The microprocessors may be dedicated to many process tasks as supervision process, communication process in addition to the considered task

Industrial Application of a Second Order Sliding Mode Observer for Speed and Flux Estimation in Sensorless Induction Motor

51

As the mechanical position and magnetic variables are unknown, $d - q$ frame is well appropriate for sensorless observer based control design.

IM **parameters:**

- R_S: Stator resistance (*Ohms*).
- R_R: Rotor resistance (*Ohms*).
- L_S: Stator inductance (*Ohms*).
- L_R: Rotor inductance (*H*).
- L_M: Mutual inductance (*H*).
- p: number of pole pairs.
- f: viscous friction coefficient (*Nm.s/rad*).
- J: inertia (*Kg.m²*).

IM **variables :**

- $v_{s\alpha,\beta}$: Stator voltage (V).
- $i_{s\alpha,\beta}$: Stator current (A).
- $\phi_{r\alpha,\beta}$: Rotor flux (Wb).
- Ω: Mechanical speed (*rad/s*).
- T_l: Load torque (Nm).

In order to construct the proposed observer for an industrial application, we work with a per unit model, under the following equations :

$$
\begin{cases}
\dot{x}_1 = -\gamma x_1 + \theta \left(b x_3 + c x_5 x_4 \right) + \xi v_1 \\
\dot{x}_2 = -\gamma x_2 + \theta \left(b x_4 - c x_5 x_3 \right) + \xi v_2 \\
\dot{x}_3 = a x_1 - b x_3 - c x_5 x_4 \\
\dot{x}_4 = a x_2 - b x_4 + c x_5 x_3 \\
\dot{x}_5 = h \left(x_3 x_2 - x_4 x_1 \right) - d x_5 - e T_l
\end{cases}
\tag{2}
$$

With the following parameters:

$$
x_1 = \frac{i_{s\alpha}}{I_{ref}} \qquad x_2 = \frac{i_{s\beta}}{I_{ref}} \qquad x_3 = \frac{\omega_{ref}\phi_{r\alpha}}{V_{ref}} \qquad x_4 = \frac{\omega_{ref}\phi_{r\beta}}{V_{ref}} \qquad x_5 = \frac{p\Omega}{\omega_{ref}}
$$

$$
a = \frac{M_{sr} I_{ref} \omega_{ref}}{T_r V_{ref}} \qquad b = \frac{1}{\tau_r} \qquad c = \omega_{ref} \qquad d = \frac{f_v}{J}
$$

$$
e = \frac{p}{J\omega_{ref}} \qquad h = \frac{p^2 M_{sr} I_{ref} V_{ref}}{J\omega_{ref}^2 L_r} \qquad \theta = \frac{KV_{ref}}{I_{ref}\omega_{ref}} \qquad \xi = \frac{V_{ref}}{\sigma L_s I_{ref}}
$$

$$
\sigma = 1 - \frac{M_{sr}^2}{L_s L_r} \qquad \gamma = \frac{R_s L_r^2 + R_r M_{sr}^2}{\sigma L_s L_r^2} \qquad \tau_r = \frac{L_r}{R_r} \qquad K = \frac{M_{sr}}{\sigma L_s L_r}
$$

Thus for the sake of homogeneity, hereafter experimental results will be given in per-unit (p.u.).

2.2. Observability

The *IM* observability has been studied by several authors (see for example [3], [13], [9]). In [9], it is proved that the IM observability cannot be established in the particular case when fluxes $\Phi_{r\alpha}$, $\Phi_{r\beta}$ and speed Ω are constant, even if we use the higher derivatives of currents. This is a sufficient and necessary condition for lost of observability.

This operating case match to the following physically interpretation:

Constant fluxes $(\dot{\phi}_{r\alpha} = \dot{\phi}_{r\beta} = 0)$

With ω_s the stator voltage pulsation and T_{em} the electromagnetic torque.

$$\omega_s = p\Omega + \frac{R_r T_{em}}{p\Phi_{rd}^2} = 0 \tag{3}$$

where $\Phi_{rd}^2 = \phi_{r\alpha}^2 + \phi_{r\beta}^2$ is the square of the direct flux in (d, q) frame.

Constant speed $(\dot{\Omega} = 0)$

$$T_{em} = f\Omega + T_l \tag{4}$$

Thanks to previous equations, we obtain:

$$T_l = -\left(f + \frac{p^2 \Phi_{rd}^2}{R_r}\right)\Omega \tag{5}$$

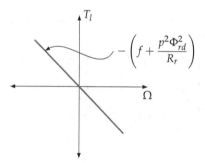

Figure 1. Inobservability curve

The unobservability curve in the map (T_l, Ω) is shown in figure (1).

Obviously, the observability is lost gradually when we approach this curve [9].

3. Second order sliding mode observer

3.1. Super twisting algorithm: An academic example

Sliding modes were used at first, as a control technique, but in the recent years it presented as a very good tool for observer design [17], [17], [5].

Industrial Application of a Second Order Sliding Mode Observer for Speed and Flux Estimation in
Sensorless Induction Motor

53

Considering the following system:

$$\begin{cases} \dot{x}_1 = x_2 \\ \dot{x}_2 = f(x,t) \\ y = h(x) = x_1 \end{cases} \tag{6}$$

With $f(x,t)$ a bounded function.

For system (6), a second order sliding mode observer is designed in the following way:

$$\begin{cases} \dot{\hat{x}}_1 = \hat{x}_2 + \lambda |e_1|^{\frac{1}{2}} sign(e_1) \\ \dot{\hat{x}}_2 = \alpha sign(e_1) \end{cases} \tag{7}$$

With $\lambda, \alpha > 0$ and $e_1 = x_1 - \hat{x}_1$.

The efficiency of the this strategy depends on coefficients α and λ. For second order system (6) we show convergence of estimated variables $(\hat{x}_1, \hat{x}_2)$ to (x_1, x_2) by studying dynamics errors $\dot{e}_1$ and $\dot{e}_2$.

Thus

$$\begin{cases} \dot{e}_1 = \dot{x}_1 - \dot{\hat{x}}_1 = e_2 - \lambda |x_1 - \hat{x}_1|^{\frac{1}{2}} sign(x_1 - \hat{x}_1) \\ \dot{e}_2 = \dot{x}_2 - \dot{\hat{x}}_2 = f(x,t) - \alpha sign(x_1 - \hat{x}_1) \end{cases} \tag{8}$$

With

$$f(x,t) \in [-f^+, f^+], \quad e_2 = x_2 - \hat{x}_2$$

And

$$\ddot{e}_1 = f(x,t) - \alpha sign(e_1) - \frac{1}{2}\lambda |e_1|^{-\frac{1}{2}} \dot{e}_1$$

Thus

$$\ddot{e}_1 \in [-f^+, f^+] - \alpha sign(e_1) - \frac{1}{2}\lambda |e_1|^{-\frac{1}{2}} \dot{e}_1$$

Where

$$f^+ = max\{f(x,t)\}$$

Conditions on λ and α that permit a convergence in finite time of $(\dot{e}_1, e_1)$ to $(0,0)$ are derived hereafter according to figure 2.

Proposition: For any initial conditions $x(0), \hat{x}(0)$, there exists a choice of λ and α such that the error dynamics $\dot{e}_1$ and $\dot{e}_2$ converge to zero in finite time and by consequence $\hat{x}_1 \longmapsto x_1$ and $\hat{x}_2 \longmapsto x_2$.

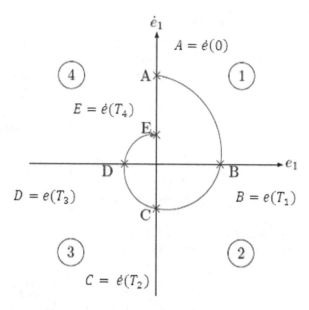

Figure 2. Upper bound of finite time convergence curve.

Proof: Consider system (6). To show the convergence of $(\hat{x}_1, \hat{x}_2)$ to (x_1, x_2) (ie., $(e_1, e_2) \rightarrow (0,0)$), we need to show that

$$\frac{|\dot{e}_1(T_2)|}{|\dot{e}_1(0)|} < 1 \tag{9}$$

of figure 2, where $\dot{e}_1(T_2) = C$ and $\dot{e}_1(0) = A$.

Figure 2 illustrates the finite time convergence behavior of the proposed observer for system 6. In what follows we will give the error trajectory for each quadrant in the worst cases.

Let consider the system's dynamic $\ddot{e}_1$

$$\ddot{e}_1 = f(x, t) - \alpha sign(e_1) - \frac{\lambda}{2}|e_1|^{-\frac{1}{2}}\dot{e}_1 \tag{10}$$

with $\dfrac{d\,|x|}{dt} = \dot{x}sign(x)$.

Equation (10) leads to

$$\ddot{e}_1 \in [-f^+, f^+] - \alpha sign(e_1) - \frac{\lambda}{2}|e_1|^{-\frac{1}{2}}\dot{e}_1 \tag{11}$$

where

$$f^+ = max(f(t, x)),$$

Industrial Application of a Second Order Sliding Mode Observer for Speed and Flux Estimation in
Sensorless Induction Motor

55

First quadrant: $e_1 > 0$ and $\dot{e}_1 > 0$

Starting from point A of figure 2 the trajectory of $\dot{e}_1 = f(e_1)$ is in the first quadrant $e_1 \geq 0$ and $\dot{e}_1 \geq 0$. The rising trajectory is given by $\ddot{e}_1 = -(\alpha - f^+)$.

By choosing

$$\alpha > f^+ \tag{12}$$

we ensure that $\ddot{e}_1 < 0$ and hence $\dot{e}_1$ decreases and tends towards the y-axis, corresponding to $\dot{e}_1 = 0$ (point B in figure 2).

Computing of $e_1(T_1)$

From (10), we have

$$\ddot{e}_1 = -(\alpha - f^+)$$

Which implies that

$$\dot{e}_1(t) = -(\alpha - f^+)t + \dot{e}_1(0) \tag{13}$$

And

$$e_1(t) = -(\alpha - f^+)\frac{t^2}{2} + \dot{e}_1(0)t$$

From (13), since $\dot{e}_1(T_1) = 0$, we obtain the necessary time for going from A to B with $B = e_1(T_1)$.

$$T_1 = \frac{\dot{e}_1(0)}{(\alpha - f^+)} \tag{14}$$

Then we can compute $e(T_1)$ as follows

$$e_1(T_1) = -(\alpha - f^+)\frac{\dot{e}_1^2(0)}{2(\alpha - f^+)} + \frac{\ddot{e}_1^2(0)}{(\alpha - f^+)}$$

$$= \frac{\dot{e}_1^2(0)}{2(\alpha - f^+)} \tag{15}$$

Second quadrant: $e_1 > 0$ and $\dot{e}_1 < 0$

In this case, $\ddot{e}_1 = -f^+ - \alpha sign(e_1) - \frac{\lambda}{2}|e_1|^{-\frac{1}{2}}\dot{e}_1$

becomes negative ($\ddot{e}_1 < 0$) on making a good choice of α which leads to

$$(\alpha + f_1^+) > -\frac{\lambda_1}{2}|e_1|^{-\frac{1}{2}}\dot{e}_1 \tag{16}$$

Since $\dot{e}_1$ is negative, then

$$|\dot{e}_1(t)| \leq \frac{2(\alpha + f_1^+)}{\lambda}|e_1(t)|^{\frac{1}{2}} \tag{17}$$

Considering by the sake of simplicity (17), $e_1 > 0$ and $\dot{e}_1 < 0$.

Integrating (17) with $e_1(0) = 0$ gives

$$\sqrt{e_1(t)} = \frac{(\alpha + f^+)}{\lambda} t \qquad (18)$$

At $t = T_2$, we should make the inverse of function (18) from point B to C in figure 2. This leads to

$$e_1(T_2) = e_1(T_1). \qquad (19)$$

Then (18) becomes

$$\sqrt{e_1(T_1)} = \frac{(\alpha + f^+)}{\lambda} T_2 \qquad (20)$$

By replacing $e_1(T_1)$ coming from (15) in equation (20), we get the necessary time for going from B to C

$$T_2 = \frac{\lambda}{(\alpha + f^+)} \frac{\dot{e}_1(0)}{\sqrt{2(\alpha - f^+)}} \qquad (21)$$

After that, by using the argument of (19) in equation (17) evaluated at $t = T_2$ in the worth case, we get

$$|\dot{e}_1(T_2)| = \frac{2(\alpha + f_1^+)}{\lambda} |e_1(T_1)|^{\frac{1}{2}} \qquad (22)$$

By replacing $e_1(T_1)$ by its expression given by (15) in (22), we get

$$|\dot{e}_1(T_2)| = \frac{2(\alpha + f_1^+)}{\lambda} \frac{|\dot{e}_1(0)|}{\sqrt{2}\sqrt{(\alpha - f^+)}} \qquad (23)$$

Thus, by satisfying inequality (9) in equation (23) λ should be chosen as

$$\lambda > (\alpha + f_1^+) \sqrt{\frac{2}{(\alpha - f^+)}} \qquad (24)$$

Finally, conditions (12) and (24) of the observer parameters are sufficient conditions guaranteeing the state convergence (i.e. the states $(e_1, \dot{e}_1)$ tend towards $e_1 = \dot{e}_1 = 0$ (Figure 2).

This ends the proof.

Moreover the convergence is in finite time, because from (14) and 21 we obtain

$$T_\infty \le \left(\sum_{i=0}^{+\infty} \left(\frac{\sqrt{2}(\alpha + f^+)}{\lambda\sqrt{\alpha - f^+}} \right)^i \right) \left(\frac{\lambda}{\sqrt{2}(\alpha + f^+)} + 1 \right) \frac{\dot{e}_1(0)}{(\alpha - f^+)}$$

as

$$\left| \frac{\sqrt{2}(\alpha + f^+)}{\lambda\sqrt{\alpha - f^+}} \right| < 1$$

Industrial Application of a Second Order Sliding Mode Observer for Speed and Flux Estimation in
Sensorless Induction Motor

57

we obtain a bounded limit

$$T_\infty \le \left(\frac{1}{1 - \frac{\sqrt{2(\alpha+f^+)}}{\lambda\sqrt{\alpha-f^+}}}\right)\left(\frac{\lambda}{\sqrt{2}(\alpha + f^+)} + 1\right)\frac{\dot{e}_1(0)}{(\alpha - f^+)}$$

Here we give simulations of a very simple example. The function $f(t, x)$ in system (6) is set equal to $sin(t)$ with $f^+ = max\{sin(t)\} = 1$. We get

$$\begin{cases} \dot{x}_1 = x_2 \\ \dot{x}_2 = sin(t) \\ y = x_1 \end{cases} \tag{25}$$

The associated observer is:

$$\begin{cases} \dot{\hat{x}}_1 = \tilde{x}_2 + \lambda|e_1|^{\frac{1}{2}}sign(e_1) \\ \dot{\tilde{x}}_2 = \alpha sign(e_1) \end{cases} \tag{26}$$

The simulation results are shown in figure 3. It can be seen that figure (3) spotlight *two* steps into Super Twisting Algorithm, which are convergence step in finite time, and sliding mode. Indeed observer is working on $t = 1s$ with $\hat{x}_1(0) = 1$ and $\tilde{x}_2(0) = 1$. $\hat{x}_1$ converges under $1s$ to x_1, and then slides along x_1 path, and equal to $\tilde{x}_2$.

3.2. Application to *Induction Motor*

At first, due to the nonlinearity of flux and speed product, the *IM* model (2) is not written in a suitable form allowing to apply the super twisting algorithm presented in previous section. To overcome this difficulty, we make the following change of variables in order to rewrite the *IM* model (2) (without $\dot{x}_5$ equation) into a form 6:

$$\begin{cases} z_1 = x_1 \\ z_2 = x_2 \\ z_3 = b x_3 + c x_5 x_4 \\ z_4 = b x_4 - c x_5 x_3 \\ z_5 = \dot{z}_3 \\ z_6 = \dot{z}_4 \end{cases} \tag{27}$$

Equation 27 is not a diffeomorphism, not an homeomorphism but only an immersion, because the dimension of x is 5 and the dimension of z is 6. Nevertheless, this immersion is used in order to avoid some singularities in a speed estimation as this will be pointed out in the next.

From the *IM* model (2) and (27), we obtain a new dynamical system as following:

$$\begin{cases} \dot{z}_1 = -z_1 + \theta z_3 + \xi v_1 \\ \dot{z}_2 = -\gamma z_2 + \theta z_4 + \xi v_2 \\ \dot{z}_3 = z_5 \\ \dot{z}_4 = z_6 \end{cases} \tag{28}$$

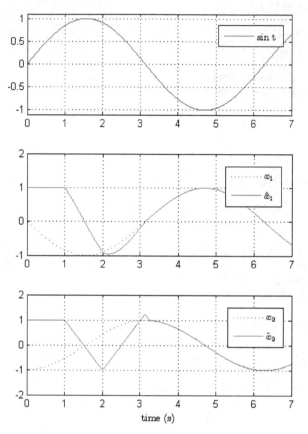

Figure 3. Super Twisting Algorithm example.

Thus, we can propose a new observer structure for dynamical system (28):

$$\begin{cases} \dot{\hat{z}}_1 = \theta\,\tilde{z}_3 - \gamma\,z_1 + \xi\,v_1 + \lambda_1\,|e_1|^{\frac{1}{2}}\,sign(e_1) \\ \dot{\hat{z}}_3 = \alpha_1\,sign(e_1) \\ \dot{\hat{z}}_2 = \theta\,\tilde{z}_4 - \gamma\,z_2 + \xi\,v_2 + \lambda_2\,|e_2|^{\frac{1}{2}}\,sign(e_2) \\ \dot{\hat{z}}_4 = \alpha_2\,sign(e_2) \\ \dot{\hat{z}}_3 = E_1\,E_2\,(\tilde{z}_5 + \lambda_3\,|e_3|^{\frac{1}{2}}\,sign(e_3)) \\ \dot{\hat{z}}_5 = E_1\,E_2\,\alpha_3\,sign(e_3) \\ \dot{\hat{z}}_4 = E_1\,E_2\,(\tilde{z}_6 + \lambda_4\,|e_4|^{\frac{1}{2}}\,sign(e_4)) \\ \dot{\hat{z}}_6 = E_1\,E_2\,\alpha_4\,sign(e_4) \end{cases} \tag{29}$$

$$\text{with } E_i \begin{cases} 1 \text{ if } e_i = z_i - \hat{z}_i = 0, i = 1,2 \\ 0 \text{ if not} \end{cases} \tag{30}$$

This observer structure depends on Super Twisting Algorithm presented in previous section and Step by Step proficiencies [7]. We propose to put in multiples-series observers with functions (E_i).

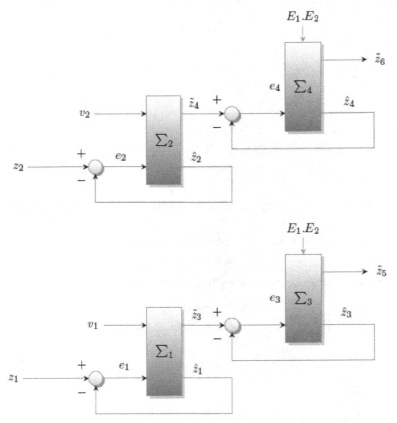

Figure 4. General *IM* Observer Structure.

The functions E_i ensure that the next steps errors do not escape too far before one has the convergent of the last step error.

The gains α_i, λ_i are chosen which respect to the reachability condition of the Super Twisting algorithm as stated in inequalities (12) and (24) of previous section. By choosing

$$\alpha_1 > max(\theta z_3), \quad \lambda_1 > (max(\theta z_3) + \alpha_1)\sqrt{\frac{2}{\alpha_1 - max(\theta z_3)}} \tag{31}$$

$$\alpha_2 > max(\theta z_4), \quad \lambda_2 > (max(\theta z_4) + \alpha_2)\sqrt{\frac{2}{\alpha_2 - max(\theta z_4)}} \tag{32}$$

$$\alpha_3 > max(z_5), \quad \lambda_3 > (max(z_5) + \alpha_3)\sqrt{\frac{2}{\alpha_3 - max(z_5)}} \tag{33}$$

$$\alpha_4 > max(z_6), \quad \lambda_4 > (max(z_6) + \alpha_4)\sqrt{\frac{2}{\alpha_4 - max(z_6)}} \tag{34}$$

and we get

$$e_1 = e_2 = e3 = e_4 = 0$$

i.e.

$$\hat{z}_1 = z_1, \quad \hat{z}_2 = z_2$$
$$\hat{z}_3 = \tilde{z}_3, \quad \hat{z}_4 = \tilde{z}_4$$
$$\tilde{z}_5 = z_5, \quad \tilde{z}_6 = z_6$$

Consequently all variables z_1, z_2, $\hat{z}_3$, $\hat{z}_4$, $\tilde{z}_5$, $\tilde{z}_6$ are available and then we can deduce *IM* variables.

We propose to treat this problem in two different cases: $\dot{x}_5 \neq 0$, and $\dot{x}_5 = 0$

CASE A : $\dot{x}_5 \neq 0$

Firstly we propose to express fluxes x_3 and x_4, from equation (27) we obtain:

$$\hat{z}_3 = bx_3 + cx_5x_4$$
$$\hat{z}_4 = bx_4 - cx_5x_3$$

We deduce

$$x_3 = \frac{\hat{z}_3 - cx_5x_4}{b} \tag{35}$$

$$x_4 = \frac{\hat{z}_4 + cx_5x_3}{b} \tag{36}$$

By substituting x_4 by its expression in (35) and x_3 in (36) we have:

$$x_3 = \frac{\hat{z}_4 + \frac{c}{b}z_3\,x_5}{b + \frac{c^2\,x_5^2}{b}}$$
$$x_4 = \frac{\hat{z}_3 - \frac{c}{b}z_4\,x_5}{b + \frac{c^2\,x_5^2}{b}} \tag{37}$$

Now let us express x_5. From (2) we know

$$\dot{x}_3 = ax_1 - bx_3 - cx_5x_4$$
$$\dot{x}_4 = ax_2 - bx_4 + cx_5x$$

Industrial Application of a Second Order Sliding Mode Observer for Speed and Flux Estimation in Sensorless Induction Motor

61

Firstly we propose to write $\dot{x}_3$ and $\dot{x}_4$ as a function of variables z. By using (27) in (38), we get:

$$\dot{x}_3 = az_1 - \hat{z}_3$$
$$\dot{x}_4 = az_2 - \hat{z}_4 \tag{39}$$

By Replacing (39) in (38) and using the two first equations in (27), it follows

$$\hat{z}_3 = bx_3 + cx_5x_4 \tag{40}$$
$$\hat{z}_4 = bx_4 - cx_5x_3 \tag{41}$$

Taking the time derivative of (41) and using third-fourth equations in (27) yields to

$$\dot{z}_3 = \tilde{z}_5 = b\dot{x}_3 + c\dot{x}_5x_4 + c\dot{x}_4x_5 \tag{42}$$
$$\dot{z}_4 = \tilde{z}_6 = b\dot{x}_4 - c\dot{x}_5x_3 - c\dot{x}_3x_5 \tag{43}$$

From (42) we have:

$$\dot{x}_5 = \frac{\tilde{z}_5 - b\dot{x}_3 - cx_5\dot{x}_4}{cx_4} \tag{44}$$

By substituting (44) in (43), we get

$$\tilde{z}_6 = b\dot{x}_4 - cx_5\dot{x}_3 - x_3\frac{\tilde{z}_5 - b\dot{x}_3 - cx_5\dot{x}_4}{x_4} \tag{45}$$

Then we can deduce the motor speed x_5 by replacing in (45) expressions of x_3-x_4 and $\dot{x}_3$-$\dot{x}_4$ coming from (37) and (39), respectively.

After a straightforward computations, we obtain a second order expression of x_5:

$$\pi_1 x_5^2 + \pi_2 x_5 + \pi_3 = 0 \tag{46}$$

where

$$\pi_1 = \frac{c}{b}\left[(az_2 - \hat{z}_4)\hat{z}_3 - (az_1 - \hat{z}_3)\hat{z}_4\right]$$
$$\pi_2 = \frac{c}{b}\left[b(az_1 - \hat{z}_3)\hat{z}_3 - b(az_2 - \hat{z}_4)\hat{z}_4 - \tilde{z}_5\hat{z}_3 + \hat{z}_4\tilde{z}_6\right]$$
$$\pi_3 = \hat{z}_3\left[-\tilde{z}_6 + b(az_2 - \hat{z}_4)\right] - \hat{z}_4\left[\tilde{z}_5 - b(az_1 - \hat{z}_3)\right]$$

CASE B : $\dot{x}_5 = 0$

We propose this hypothesis because of dynamical gap evolution between electrical and mechanical variables, in fact speed evolves much more slowly than currents or fluxes.

Thus with this hypothesis we simplify (42) and (43), and obtain two expressions of x_5 :

$$x_5 = \frac{\tilde{z}_5 - b\dot{x}_3}{c\dot{x}_4} \tag{47}$$

or

$$x_5 = \frac{b\dot{x}_4 - \tilde{z}_6}{c\dot{x}_3} \tag{48}$$

we change $\dot{x}_3$ by expression (35) and $\dot{x}_4$ by expression(36)

$$x_5 = \frac{\dot{z}_5 - baz_1 + b\hat{z}_3}{caz_2 - c\hat{z}_4} \tag{49}$$

or

$$x_5 = \frac{\dot{z}_6 + baz_2 - b\hat{z}_4}{caz_1 - c\hat{z}_3} \tag{50}$$

Equations (49) and (50) are true only if :

$$caz_2 - c\hat{z}_4 \neq 0 \text{ for (49) and } caz_1 - c\hat{z}_3 \neq 0 \text{ for (50)}$$

Speed estimation

In order to avoid singularities of speed estimation in (49) and (50), we use the fact that (49) and (50) are in quadrature and thus we get the estimation of speed x_5 as follows :

$$x_5 = \frac{(\dot{z}_5 - baz_1 + b\hat{z}_3)(caz_2 - c\hat{z}_4) + (\dot{z}_6 + baz_2 - b\hat{z}_4)(caz_1 - c\hat{z}_3)}{(caz_2 - c\hat{z}_4)^2 + (caz_1 - c\hat{z}_3)^2} \tag{51}$$

Flux estimation

The rotor flux are obtained by replacing the estimation speed (51) in (37)

Flux position estimation

Having the rotor flux estimation (37), we can obtain rotor flux position ρ

$$\rho = atan(\frac{x_4}{x_3}) \tag{52}$$

4. Discrete time implementation

4.1. Explicit Euler method

For the industrial application in real time, the discrete time observer is designed. The explicit Euler's method is chosen to transform continuous observer to discrete observer. This is due to the simplicity of computation. Considering a differential equation :

$$\dot{x} = f(x)$$

The explicit Euler's method with a sampling time T_e gives:

$$x(k) = x(k-1) + T_e f(x(k-1))$$

the data acquisition period T_e is also the computation period.

Applying the explicit Euler's method for the second order sliding mode observer, the discrete observer is obtained:

Industrial Application of a Second Order Sliding Mode Observer for Speed and Flux Estimation in
Sensorless Induction Motor

63

$$
\begin{cases}
\hat{z}_1(k) = \hat{z}_1(k-1) + T_e \left[\theta\, \hat{z}_3(k-1) - \gamma\, z_1(k-1) + \xi\, v_1(k-1) \right. \\
\qquad\qquad \left. + \lambda_1\, |e_1(k-1)|^{\frac{1}{2}} \, sign(e_1(k-1)) \right] \\
\hat{z}_3(k) = \hat{z}_3(k-1) + T_e\ \alpha_1\, sign(e_1(k-1)) \\
\hat{z}_2(k) = \hat{z}_2(k-1) + T_e \left[\theta\, \hat{z}_4(k-1) - \gamma\, z_2(k-1) + \xi\, v_2(k-1) \right. \\
\qquad\qquad \left. + \lambda_2\, |e_2(k-1)|^{\frac{1}{2}} \, sign(e_2(k-1)) \right] \\
\hat{z}_4(k) = \hat{z}_4(k-1) + T_e\, \alpha_2\, sign(e_2(k-1)) \\
\hat{z}_3(k) = \hat{z}_3(k-1) + T_e\, E_1\, E_2 \left[\hat{z}_5(k-1) + \lambda_3\, |e_3(k-1)|^{\frac{1}{2}}\, sign(e_3(k-1)) \right] \\
\hat{z}_5(k) = \hat{z}_5(k-1) + T_e\, E_1\, E_2\, \alpha_3\, sign(e_3(k-1)) \\
\hat{z}_4(k) = \hat{z}_4(k-1) + T_e\, E_1\, E_2 \left[\hat{z}_6(k-1) + \lambda_4\, |e_4(k-1)|^{\frac{1}{2}}\, sign(e_4(k-1)) \right] \\
\hat{z}_6(k) = \hat{z}_6(k-1) + T_e\, E_1\, E_2\, \alpha_4\, sign(e_4(k-1))
\end{cases}
\tag{53}
$$

4.2. Oversampling

To achieve good accuracy, a small sample period and fast DSP are needed. In the industrial application, the DSP clock frequency is only $150 MHz$, which does not allow a small enough sample period. So in experimentation an over-sample technique is proposed. In the following paragraphs we show that, under a few low restrictive conditions, it is possible to reduce the error of Euler's method, seen in the previous subsection (4.1).

Hereafter we first present the oversampling method in a very simple use, where $f \in C^\infty$. Assume a continuous autonomous system of the form:

$$
\dot{x} = f(x); \quad x(t_0) = x_0
\tag{54}
$$

Assume in addition that the system is discretized at a sampling time T_e. Then the system ((54)) can be approximated by the explicit Euler's method:

$$
x(t_{k+1}) = x(t_k) + T_e\, \dot{x}(t_k) + O(T_e^2)
\tag{55}
$$

For small values of T_e, $O(T_e^3)$ is neglected and the truncation error is approximately proportional to T_e^2.

Suppose now that the system (54) is discretized at two different sample rates resulting in two discrete time systems: H_1 sampled at frequency $f_{s1} = \dfrac{1}{T_e}$ and H_2 sampled at frequency $f_{s2} = \dfrac{N}{T_e}$; and let us compare the truncation error of each one after T_e seconds for N large enough. The discrete time is given by $t_{H1} = nT_e$ for H_1 and $t_{H2} = k\dfrac{T_e}{N}$ for H_2, with $n, k \in \mathbb{N}$. Assume that the initial times and the initial conditions are the same for both of them, that is: $nT_e = k\dfrac{T}{T_e} = t_0$ and $x(nT_e) = x\left(k\dfrac{T_e}{N}\right) = x(t_0) = x_0$, The dynamics of the discrete time system H_1 can be written as:

$$x((n+1)T_e) = x(nT_e) + T_e f(x(nT_e)) + O(T_e^2) \tag{56}$$

H_2, sampled at $f_{s2} = \dfrac{N}{T_e}$, evolves as follows:

$$x((k+1)\frac{T_e}{N}) = x(k\frac{T_e}{N}) + \frac{T_e}{N}f\left(x(k\frac{T_e}{N})\right) + O\left(\frac{T_e}{N}\right)^2$$

$$x((k+2)\frac{T_e}{N}) = x((k+1)\frac{T_e}{N}) + \frac{T_e}{N}f\left(x((k+1)\frac{T_e}{N})\right) + O(\frac{T_e}{N})^2$$

$$= x(k\frac{T_e}{N}) + \frac{T_e}{N}f\left(x(k\frac{T_e}{N})\right) \tag{57}$$

$$+ \frac{T_e}{N}f\left(x(k\frac{T_e}{N}) + \frac{T_e}{N}f\left(x(k\frac{T_e}{N})\right) + O\left(\frac{T_e}{N}\right)^2\right) + 2O\left(\frac{T_e}{N}\right)^2$$

For N large enough we can consider the influence of the error term $O\left(\dfrac{T_e}{N}\right)^2$ over the function $\dfrac{T_e}{N}f(\cdot)$ as a term in $O\left(\dfrac{T_e}{N}\right)^3$, then:

$$x((k+2)\frac{T_e}{N}) \approx x(k\frac{T_e}{N}) + \frac{T_e}{N}f\left(x(k\frac{T_e}{N})\right) + \frac{T_e}{N}f\left\{x\left(k\frac{T_e}{N}\right) + \frac{T_e}{N}f\left(x\left(k\frac{T_e}{N}\right)\right)\right\}$$

$$+ 2O\left(\frac{T_e}{N}\right)^2 \tag{58}$$

$$= x(k\frac{T_e}{N}) + \frac{T_e}{N}f\left(x(k\frac{T_e}{N})\right) + \frac{T_e}{N}f(x((k+1)\frac{T_e}{N})) + 2O\left(\frac{T_e}{N}\right)^2$$

So, in a general way, we have:

$$x((k+N)\frac{T_e}{N}) \approx x(k\frac{T_e}{N}) + \frac{T_e}{N}\sum_{i=k}^{k+N-1} f(x(i\frac{T_e}{N})) + NO\left(\frac{T_e}{N}\right)^2 \tag{59}$$

As we can see from (56) and (59), the truncation errors of the discrete systems H_1 and H_2 are

$$\varepsilon_1 = O\left(T_e^2\right) \tag{60}$$

and

$$\varepsilon_2 = NO\left(\left(\frac{T_e}{N}\right)^2\right) \tag{61}$$

The truncation errors ε_1 and ε_2, given by (60) and (61) respectively give

$$NO\left(\left(\frac{T_e}{N}\right)^2\right) \approx \frac{O\left(T_e^2\right)}{N} \tag{62}$$

Industrial Application of a Second Order Sliding Mode Observer for Speed and Flux Estimation in
Sensorless Induction Motor

65

$$\varepsilon_2 \approx \frac{\varepsilon_1}{N} \tag{63}$$

Then the oversampled system H_2 reduces the truncation error about N times.

In practice, to achieve the benefits of oversampling, we emulate this technique based on the assumption that between two consecutive samples of an input signal, its derivative is nearly constant. In this way the new" samples are obtained by linear interpolation between consecutive "measured" samples. This technique is shown in figure 5. On the top the

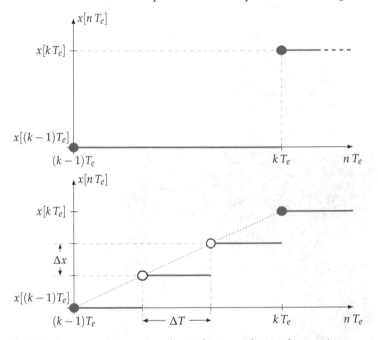

Figure 5. Comparison between sampling and oversampling implementation.

"classical sampling" is shown and on the bottom, the oversampling technique is depicted. As we can see, the sample period is reduced three times its original value, that is, $\frac{T_e}{N} = \frac{T_e}{3}$. This technique reduces the truncation error, inherent to Euler's method, three times. The benefits of this technique are exposed and validated by experimental tests.

5. Experimentations

5.1. Test bench

Table (1) presents all electrical and mechanical parameters of Induction machine used in practice, and in Table (2) main VAR-CNTRL card features are presented.

The tunning parameters $\alpha_i, \lambda_i, i = 1, ..., 4$ of the proposed observer are chosen according to inequalities (31), 32, 33 and 34 to satisfy convergence conditions.

P_N	rated power	1.5KW
V_N	rated voltage	230V
I_N	rated current	3.2A
F_N	rated frequency	50Hz
N_N	rated speed	2998tr/min
p	number of pair of poles	1
R_S	stator resistance	4.2Ω
R_R	rotor resistance	2.8Ω
L_S	stator inductance	0.522 H
L_R	stator inductance	0.537 H
M_{SR}	mutual inductance	0.502 H
f	viscous coefficient	1N.s/rad
α_1, λ_1	tunning parameters	$\alpha_1 = 1500, \lambda_1 = 2500$
α_2, λ_2	tunning parameters	$\alpha_2 = \alpha_1, \lambda_2 = \lambda_1$
α_3, λ_3	tunning parameters	$\alpha_3 = 1500, \lambda_3 = 2000$
α_4, λ_4	tunning parameters	$\alpha_4 = \alpha_3, \lambda_4 = \lambda_3$

Table 1. Induction machine and observer parameters.

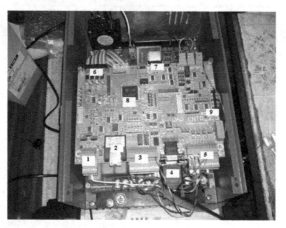

Figure 6. VAR-CNTRL card a product of GS Maintenance.

1	Analog Input/Output connectors. (3 Inputs /3 0utputs)
2	Communication port. (1 $RS232$)
3	Logical Output connector. (6 Outputs)
4	QEP connector. (A-B-Z)
5	Logical Input connector. (8 Isolated Inputs)
6	Supply voltage connector. (3.3V- 5V - ($\pm$15V) - 24V)
7	Measurements connector. (V_{DC}, I_A, I_B)
8	DSP $TMS320F2812$
9	PWM connector. (6 Output signals).

Table 2. VAR-CNTRL card main elements.

VAR-CNTRL is a electronic card designed by **GS Maintenance** and dedicated to motor control (Synchronous, Induction machine, Brushless, and DC motor). Equipped with a **DSP TMS320F2812** from **Texas Instrument**,this component is a fixed point; data are represented under 32 bits.

Practicals tests have been done under the following configurations:

- Fe , Sampling frequency of 8KHZ.
- Fcyc, **DSP** clock frequency of 150MHZ.
- 1024 points encoder, as speed sensor.
- **ADC's** (Analog-to-Digital Coder) of 12 bits provide bus voltage (V_{DC}), and phase currents (I_A, I_B) frames under 12 bits.

In addition to the **VAR-CNTRL**, a **MMI** (Man Machine Interface) permits to visualize **DSP** data registers in representation format 8.8 that means possible variations are from [-127.996 to 128].

To summarize our Bed Test description, we have :

- An **IM**.
- A two-level **VSI** (Voltage Source Inverter).
- A control card , **VAR-CNTRL**.
- A **MMI**.
- A speed sensor, a voltage sensor, and two current sensors.

5.2. Results

In this section we propose some experimentation results, that allow the following points:

- Validate Super Twisting Algorithm convergence.
- Evaluate Oversampling method efficiency.
- Evaluate Motor variables estimation.

In section 4.1 we introduced Euler Explicit Sampling Method to discretize a continuous system. Some technical limits about sampling frequency Fe lead us to introduce Oversampling strategy (c.f. section 4.2) . At first glance we propose to validate Super Twisting Observer strategy (c.f. system 7), we will take account of subsystem Σ_1 in figure 4 with the following entries : (v_1 , z_1), and outputs : ($\hat{z}_1$, $\hat{z}_3$).

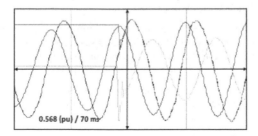

0.568 (pu) / 70 ms

Figure 7. MMI capture : (z_1,v_1) and (,) on convergence phase.

On figure 7 we validate the convergence of Σ_1 in figure 4, we can see that under some initials values $\hat{z}_1$ converge to z_1 in a finite time.

Figures 8 and 9 permit to assume that oversampling method is efficient, in fact we see that signals estimated by the observer (53) of the subsystem Σ_1 in figure 4 are much more better with an oversampling than without.

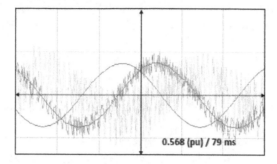

Figure 8. MMI capture : (z_1, v_1) and (,) without oversampling method.

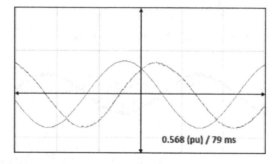

Figure 9. MMI capture : (z_1, v_1) and (,) with oversampling method. ($N = 10$.)

Thus at the same operating point, we assume that with oversampling method we improve efficiency of the algorithm. With this validated data we can now abort estimation of *IM* magnetic (x_3, x_4) and mechanical (x_5) variables including the rotor position flux ρ given in equations 51, (35-36) and (52) respectively.

The main objective of this work is to provide a motor speed estimation without any mechanical sensor, and then drive it. Note that the speed sensor is only used in comparison of estimated speed with its measure. To validate our strategy we propose some tests into different conditions.

Figures 12 and 13 permit to validate accuracy of estimated speed compare to measured speed in high variation range. However, it is admit that at low and very low speed, estimated speed damages more and more, as we can see on figures 14 and 15.

Now we propose some dynamical test results. During acceleration and deceleration phases (c.f. 16), estimated speed is steel working although there is small delay between x_5 and $\hat{x}_5$.

Industrial Application of a Second Order Sliding Mode Observer for Speed and Flux Estimation in
Sensorless Induction Motor

69

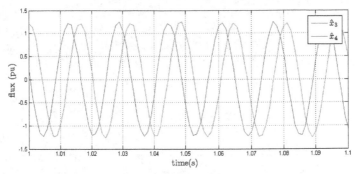

Figure 10. Flux Estimation: x_3 and x_4 during static phase.

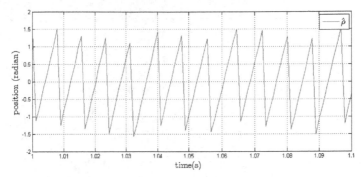

Figure 11. Estimated of rotor flux position: ρ.

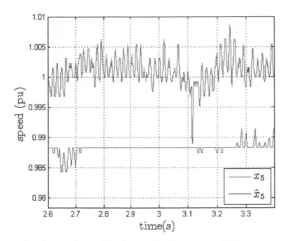

Figure 12. Measure and estimate of speed during static phase: x_5 and $\hat{x}_5$

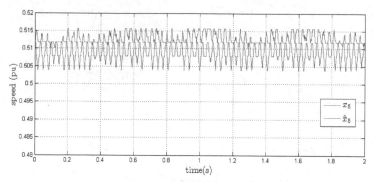

Figure 13. Measure and estimate of speed during static phase: x_5 and $\hat{x}_5$

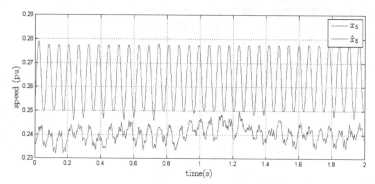

Figure 14. Measure and estimate of speed during static phase: x_5 and $\hat{x}_5$

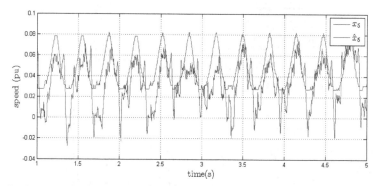

Figure 15. Measure and estimate of speed during static phase: x_5 and $\hat{x}_5$

During static phase operation we saw that at low and very low speed , speed observation does not work very well. However on figures (17) and (18) we cross 0 speed, we denote a small

divergence as small as the time to cross it; in fact this phenomenon underlines that speed is
non observable with low current dynamic.

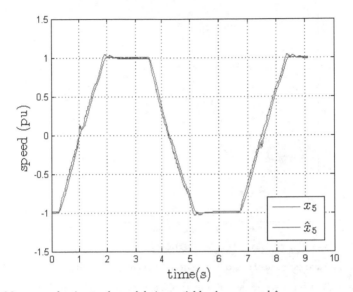

Figure 16. Measure and estimate of speed during variable phase: x_5 and $\hat{x}_5$

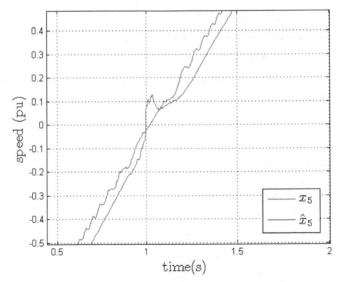

Figure 17. Measure and estimate of speed during acceleration phase: x_5 and $\hat{x}_5$

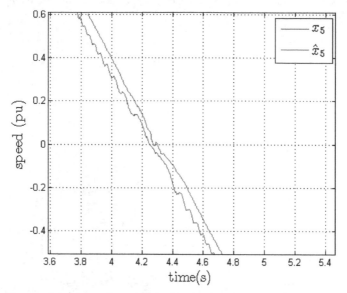

Figure 18. Measure and estimate of speed during deceleration phase: x_5 and $\hat{x}_5$

Figures 12 and 13 at high speed, show that speed approximation proposed in equation (51) work and permit to obtain magnitude and speed sign. This efficiency is also proved during dynamical phases as we can see on figure 16.

About this bad results, we have 2 arguments:

- Parameters error, mainly on stator and rotor resistance (R_S, R_R).
- Poor current dynamic, combined to digitizing error at low frequency working.

To overcome all this features, we propose to use an on-line resistor measurement of stator threw temperature.

6. Conclusion

Through this chapter an original method of observation without mechanical sensors for induction machine was introduced.

Designed for a embedded system (VAR-CNTRL) equipped with a fixed point DSP, we carried out various tests of validation.

We used concept of Sliding Mode through Super Twisting Algorithm, and oversampling method being based on the explicit Euler development. The contribution of this paper is mainly based on the applicability of the proposed observer for sensorless induction motor when a basic microprocessors are used in an industrial context.

At the time of the setting works of our strategy some technical constraints brought us to introduce a news strategy.

Industrial Application of a Second Order Sliding Mode Observer for Speed and Flux Estimation in
Sensorless Induction Motor

73

Thus the practical results permit us to do a first assessment:

- we validate our oversampling method introduced to overcome low speed data acquisition.
- we validate speed estimation during static and dynamic steps.
- we obtained an image of rotor flux(x_3, and x_4), and also rotor position.

Compared with mechanical sensor the precision provides by the observer on the size speed offer a precision inferior or equal to 5% in the operating speed range from: 25% to 100%.

In term of prospects, it possible to improve the threshold of operation in low mode (25% to 5%) by adaptation oversampling number to stator frequency value, indeed a larger sample number could improve approximation of the continuous system .

In next step some tests will done to validate:

- Validation of hardiness to load variation.
- Validation in closed loop.

About Observability loose at very low speed a first solution could be to switch with a speed estimator.

Author details

Sebastien Solvar, Malek Ghanes, Leonardo Amet, Jean-Pierre Barbot and Gaëtan Santomenna
ECS - Lab, ENSEA and GS Maintenance, France

7. References

[1] Aurora, C. & Ferrara, A. [2007]. A sliding mode observer for sensorless induction motor speed regulation, *International Journal of Systems Science* 38(11): 913–929.

[2] Bartolini, G., Ferrara, A. & Usani, E. [1998]. Chattering avoidance by second-order sliding mode control, *Automatic control, IEEE Transactions on* 43(2): 241–246.

[3] Canudas De Wit, C., Youssef, A., Barbot, J., Martin, P. & Malrait, F. [2000]. Observability conditions of induction motors at low frequencies, *Decision and Control, 2000. Proceedings of the 39th IEEE Conference on*, Vol. 3, IEEE, pp. 2044–2049.

[4] Chiasson, J. [2005]. *Modeling and high performance control of electric machines*, Vol. 24, Wiley-IEEE Press.

[5] Davila, J., Fridman, L. & Levant, A. [2005]. Second-order sliding-mode observer for mechanical systems, *IEEE Transactions on Automatic Control* Vol. 50(11): 1785–1789.

[6] Dib, A., Farza, M., MŠSaad, M., Dorléans, P. & Massieu, J. [2011]. High gain observer for sensorless induction motor, *World Congress*, Vol. 18, pp. 674–679.

[7] Floquet, T. & Barbot, J. [2007]. Super twisting algorithm-based step-by-step sliding mode observers for nonlinear systems with unknown inputs, *International Journal of Systems Science* 38(10): 803–815.

[8] Ghanes, M., Barbot, J., De Leon, J. & Glumineau, A. [2010]. A robust sensorless output feedback controller of the induction motor drives: new design and experimental validation, *International Journal of Control* 83(3): 484–497.

[9] Ghanes, M., De Leon, J. & Glumineau, A. [2006]. Observability study and observer-based interconnected form for sensorless induction motor, *Decision and Control, 2006 45th IEEE Conference on*, IEEE, pp. 1240–1245.

[10] Ghanes, M. & Zheng, G. [2009]. On sensorless induction motor drives: Sliding-mode observer and output feedback controller, *Industrial Electronics, IEEE Transactions on* 56(9): 3404–3413.

[11] Hilairet, M., Auger, F. & Berthelot, E. [2009]. Speed and rotor flux estimation of induction machines using a two-stage extended kalman filter, *Automatica* 45(8): 1819–1827.

[12] Holtz, J. [2006]. Sensorless control of induction machinesǓwith or without signal injection ?, *Industrial Electronics, IEEE Transactions on* 53(1): 7–30.

[13] Ibarra-Rojas, S., Moreno, J. & Espinosa-Pérez, G. [2004]. Global observability analysis of sensorless induction motors, *Automatica* 40(6): 1079–1085.

[14] Khalil, H., Strangas, E. & Jurkovic, S. [2009]. Speed observer and reduced nonlinear model for sensorless control of induction motors, *Control Systems Technology, IEEE Transactions on* 17(2): 327–339.

[15] Kubota, H., Matsuse, K. & Nakano, T. [1993]. Dsp-based speed adaptive flux observer of induction motor, *Industry Applications, IEEE Transactions on* 29(2): 344–348.

[16] Leppänen, V.-M. [2003]. *Low-Frequency Signal-Injection Method for Speed Sensorless Vector Control of Induction Motors*, Oxford University Press, ISBN 0-19-856465-1.

[17] Levant, A. [1998]. Robust exact differentiation via sliding mode technique, *Automatica* 34(3): 379–384.

[18] Levant, A. [2003]. Higher-order sliding modes, differentiation and output-feedback control, *International Journal of Control* 76(9-10): 924–941.

[19] Maiti, S., Verma, V., Chakraborty, C. & Hori, Y. [2012]. An adaptive speed sensorless induction motor drive with artificial neural network for stability enhancement, *IEEE Transactions on Industrial Informatics* .

[20] Perruquetti, W. & Barbot, J. [2002]. *Sliding mode control in engineering*, Vol. 11, CRC.

[21] Schauder, C. [1992]. Adaptive speed identification for vector control of induction motors without rotational transducers, *Industry applications, IEEE Transactions on* 28(5): 1054–1061.

[22] Vas, P. [1998]. *Sensorless vector and direct torque control*, Vol. 729, Oxford university press Oxford, UK.

Chapter 4

The Takagi-Sugeno Fuzzy Controller Based Direct Torque Control with Space Vector Modulation for Three-Phase Induction Motor

José Luis Azcue, Alfeu J. Sguarezi Filho and Ernesto Ruppert

Additional information is available at the end of the chapter

1. Introduction

The Direct Torque Control (DTC) has become a popular technique for three-phase Induction Motor (IM) drives because it provides a fast dynamic torque response without the use of current regulators [23][9], however, nowadays exist some other alternative DTC schemes to reduce the torque ripples using the Space Vector Modulation (SVM) technique [11][14]. In general the use of fuzzy systems does not require the accurate mathematic model of the process to be controlled. Instead, it uses the experience and knowledge of the involved professionals to construct its control rule base. Fuzzy logic is powerful in the motor control area, e.g., in [1] the PI and Fuzzy Logic Controllers (FLC) are used to control the load angle which simplifies the IM drive system.

In [8] the FLC is used to obtain the reference voltage vector dynamically in terms of torque error, stator flux error and stator flux angle. In this case both torque and stator flux ripples are remarkably reduced. In [15] the fuzzy PI speed controller has a better response for a wide range of motor speed. Different type of adaptive FLC such as self-tuning and self-organizing controllers has also been developed and implemented in[20][4].

In [18], [13] and [10] are proposed fuzzy systems which outputs are a specific voltage vector numbers, similarly to the classic DTC scheme[23]. On the other hand, in [26] is proposed a fuzzy inference system to modulate the stator voltage vector applied to the induction motor, but it consider the stator current as an additional input.

In [19] two fuzzy controllers are used to generate the two components of the reference voltage vector instead of two PI controllers, similarly, in [7] flux and torque fuzzy controllers are designed to substitute the original flux and torque PI controllers, but these schemes use two independent fuzzy controllers, one for the flux control and another one for the torque control.

Unlike the schemes mentioned before, the aim of this chapter is to design a Takagi-Sugeno (T-S) Fuzzy controller to substitute flux and torque PI controllers in a conventional DTC-SVM scheme. The T-S fuzzy controller calculates the quadrature components of the stator voltage vector represented in the stator flux reference frame. The rule base for the proposed controller is defined in function of the stator flux error and the electromagnetic torque error using trapezoidal and triangular membership functions. The direct component of the stator voltage takes a linear combination of the inputs as a consequent part of the rules, however, the quadrature component of the stator voltage takes the similar linear combination used in the first output but with the coefficients interchanged, not to be necessary another different coefficients values for this output.

The simulation results shown that the proposed T-S fuzzy controller for the DTC-SVM scheme have a good performance in terms of rise time (t_r), settling time (t_s) and torque ripple when it was tested at different operating conditions validating the proposed scheme. The chapter is organized as follows. In section 2 the direct torque control principles of the DTC for three-phase induction motor is presented. In section 3 the topology of the proposed control scheme is analyzed and in section 4 the proposed T-S fuzzy controller is described in detail mentioning different aspects of its design. Section 5 presents the simulations results of T-S fuzzy controller, and in the end, the conclusion is given in Section 6.

2. Direct Torque Control principles

2.1. Dynamical equations of the three-phase induction motor

By the definitions of the fluxes, currents and voltages space vectors, the dynamical equations of the three-phase induction motor in stationary reference frame can be put into the following mathematical form [25]:

$$\vec{u}_s = R_s\vec{i}_s + \frac{d\vec{\psi}_s}{dt} \tag{1}$$

$$0 = R_r\vec{i}_r + \frac{d\vec{\psi}_r}{dt} - j\omega_r\vec{\psi}_r \tag{2}$$

$$\vec{\psi}_s = L_s\vec{i}_s + L_m\vec{i}_r \tag{3}$$

$$\vec{\psi}_r = L_r\vec{i}_r + L_m\vec{i}_s \tag{4}$$

Where $\vec{u}_s$ is the stator voltage space vector, $\vec{i}_s$ and $\vec{i}_r$ are the stator and rotor current space vectors, respectively, $\vec{\psi}_s$ and $\vec{\psi}_r$ are the stator and rotor flux space vectors, ω_r is the rotor angular speed, R_s and R_r are the stator and rotor resistances, L_s, L_r and L_m are the stator, rotor and mutual inductance, respectively.

The electromagnetic torque is expressed in terms of the cross-vectorial product of the stator and the rotor flux space vectors.

$$t_e = \frac{3}{2}P\frac{L_m}{L_rL_s\sigma}\vec{\psi}_r \times \vec{\psi}_s \tag{5}$$

$$t_e = \frac{3}{2}P\frac{L_m}{L_rL_s\sigma}|\vec{\psi}_r||\vec{\psi}_s|\sin(\gamma)$$

Where γ is the load angle between stator and rotor flux space vector, P is a number of pole pairs and $\sigma = 1 - L_m^2/(L_s L_r)$ is the dispersion factor.

The three-phase induction motor model was implemented in MATLAB/Simulink as is shown in [3], the code source of this implementation is shared in MATLAB CENTRAL [2].

2.2. Direct Torque Control

In the direct torque control if the sample time is short enough, such that the stator voltage space vector is imposed to the motor keeping the stator flux constant at the reference value. The rotor flux will become constant because it changes slower than the stator flux. The electromagnetic torque (6) can be quickly changed by changing the angle γ in the desired direction. This angle γ can be easily changed when choosing the appropriate stator voltage space vector.

For simplicity, let us assume that the stator phase ohmic drop could be neglected in (1). Therefore $d\vec{\psi}_s/dt = \vec{u}_s$. During a short time Δt, when the voltage space vector is applied it has:

$$\Delta\vec{\psi}_s \approx \vec{u}_s \cdot \Delta t \tag{7}$$

Thus the stator flux space vector moves by $\Delta\vec{\psi}_s$ in the direction of the stator voltage space vector at a speed which is proportional to the magnitude of the stator voltage space vector. By selecting step-by-step the appropriate stator voltage vector, it is possible to change the stator flux in the required direction.

2.2.1. Stator-flux-oriented direct torque control

The stator-flux-oriented direct torque control (SFO-DTC) based on space vector modulation scheme have two PI controllers as is shown in Fig. 2. This control strategy relies on a simplified description of the stator voltage components expressed in stator-flux-oriented coordinates

$$u_{ds} = R_s i_{ds} + \frac{d\psi_s}{dt} \tag{8}$$

$$u_{qs} = R_s i_{qs} + \omega_s \psi_s \tag{9}$$

Therefore, in this reference frame the stator flux quadrature component is zero as is shown in Fig. 1 that means $\psi_s = \psi_{ds}$ and $\psi_{qs} = 0$. Also, in this reference frame the electromagnetic torque is calculated by

$$T_{em} = \frac{3P}{2}\psi_s i_{qs} \tag{10}$$

$$i_{qs} = \frac{2}{3P}\frac{T_{em}}{\psi_s} \tag{11}$$

However, if the equation (11) is substitute in the equation (9), we can obtain the expression to control the electromagnetic torque applying and appropriate stator voltage quadrature

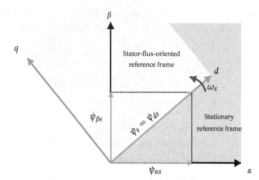

Figure 1. Stator-flux-oriented reference frame.

component, it is:

$$u_{qs} = \frac{2}{3P} R_s \frac{T_{em}}{\psi_s} + \omega_s \psi_s \tag{12}$$

From equation (8), the stator flux is controlled with the stator voltage direct component u_{ds}. For every sampled period T_s, the equation (8) is approximated by

$$u_{ds} = R_s i_{ds} + \Delta\psi_s / T_s \tag{13}$$

When the three-phase IM operates at high speeds the term $R_s i_{ds}$ can be neglected and the stator voltage can to become proportional to the stator flux change with a switching frequency $1/T_s$. However, at low speeds the term $R_s i_{ds}$ is not negligible and with the aim to correct this error is used the PI controller, it is:

$$u_{ds}^* = (K_{P\psi} + K_{I\psi}/s)(\psi_s^* - \hat{\psi}_s) \tag{14}$$

From the equation (12), the electromagnetic torque can be controlled with the stator voltage quadrature component if the term $\omega_s \psi_s$ is decoupled. A simple form to decoupled it is adding the term $\omega_s \psi_s$ to the output of the controller as is shown in Fig. 2. Then, the PI controller is used to control the electromagnetic torque, it is:

$$u_{qs}^* = (K_{PT_{em}} + K_{IT_{em}}/s)(T_{em}^* - \hat{T}_{em}) + \omega_s \psi_s \tag{15}$$

Finally, the outputs of the PI flux and PI torque controllers can be interpreted as the stator voltage components in the stator-flux-oriented coordinates [6].

Where ω_s is the angular speed of the stator flux vector. The equations (12) and (13) show that the component u_{ds} has influence only on the change of stator flux magnitude, and the component u_{qs}, if the term $\omega_s \psi_s$ is decoupled, can be used for torque adjustment. Therefore, after coordinate transformation $dq/\alpha\beta$ into the stationary reference frame, the command values u_{ds}^*, u_{qs}^* are delivered to SVM. In [3] this scheme is analyzed in detail.

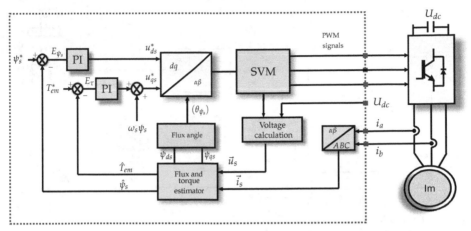

Figure 2. Conventional stator-flux-oriented direct torque control scheme.

The SFO-DTC based on space vector modulation scheme requires the flux and the torque estimators, which can be performed as it is proposed in this chapter, this scheme is used to implement the T-S fuzzy controller proposed.

3. The proposed direct torque control scheme

The Figure 3 shows the proposed DTC-SVM scheme, this scheme only needs sense the DC link and the two phases of the stator currents of the three-phase induction motor. In the DTC-SVM scheme the electromagnetic torque error (E_T) and the stator flux error (E_{ψ_s}) are the inputs and the stator voltage components are the outputs of the Takagi-Sugeno fuzzy controller, these outputs are represented in the stator flux reference frame. Details about this controller will be presented in the next section.

3.1. Stator voltage calculation

The stator voltage calculation use the DC link voltage ($\mathbf{U}_{dc}$) and the inverter switch state (S_{Wa}, S_{Wb}, S_{Wc}) of the three-phase two level inverter. The stator voltage vector $\vec{u}_s$ is determined as in [5]:

$$\vec{u}_s = \frac{2}{3}\left[\left(S_{Wa} - \frac{S_{Wb} + S_{Wc}}{2}\right) + j\frac{\sqrt{3}}{2}(S_{Wb} - S_{Wc})\right]U_{dc} \tag{16}$$

3.2. Space vector modulation technique

In this work is used the space vector modulation (SVM) technique with the aim to reduce the torque ripple and total harmonic distortion of the current, is therefore necessary to understand the operation and fundamentals that governing their behavior. This concept was discussed in publications such as [24], [12] and [27]. For our purpose the basic ideas are summarized. In Fig. 4 and Fig. 5 are shown the three-phase two level inverter diagram, where the state of the

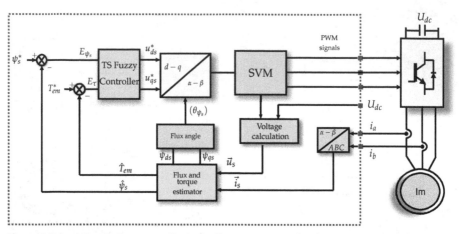

Figure 3. Takagi-Sugeno fuzzy controller in the direct torque control with space vector modulation scheme.

switches follow the following logic.

$$S_{Wi} = \begin{cases} 1, & \text{the switch } S_{Wi} \text{ is ON and the switch } \bar{S}_{Wi} \text{ is OFF} \\ 0, & \text{the switch } S_{Wi} \text{ is OFF and the switch } \bar{S}_{Wi} \text{ is ON} \end{cases} \tag{17}$$

Where i=a,b,c and considering that the switch $\bar{S}_{Wi}$ is the complement of S_{Wi} is possible to resume all the combinations only considering the top switches as is shown in Table 1.

Vector	S_{Wa}	S_{Wb}	S_{Wc}
$\vec{S}_0$	0	0	0
$\vec{S}_1$	1	0	0
$\vec{S}_2$	1	1	0
$\vec{S}_3$	0	1	0
$\vec{S}_4$	0	1	1
$\vec{S}_5$	0	0	1
$\vec{S}_6$	1	0	1
$\vec{S}_7$	1	1	1

Table 1. Switching vectors

Where $\vec{S}_0, \vec{S}_1, \vec{S}_2, \vec{S}_3, \vec{S}_4, \vec{S}_5, \vec{S}_6$ and $\vec{S}_7$ are switching vectors. These switching vectors generate six active voltage vectors ($\vec{U}_1, \vec{U}_2, \vec{U}_3, \vec{U}_4, \vec{U}_5$ and $\vec{U}_6$) and two zero voltage vectors ($\vec{U}_0$ and $\vec{U}_7$) as are shown in the Figures 6 and 7. The generalized expression to calculate the active and zero voltage vectors is:

$$\vec{U}_n = \begin{cases} \frac{2}{3}\sqrt{3}U_{dc} \cdot e^{j(2n-1)\frac{\pi}{6}} & , n = 1, ..., 6 \\ 0 & , n = 0, 7 \end{cases} \tag{18}$$

Where U_{dc} is the DC link voltage.

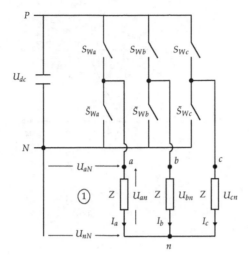

Figure 4. Three-phase two level inverter with load

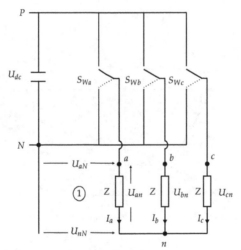

Figure 5. Simplified three-phase two level inverter with load

In Fig. 6 the hexagon is divided in six sectors, and any reference voltage vector is represented as combination of adjacent active and zero voltage vectors, e.g. the voltage vector $\vec{U}^*$ is localized in sector I between active vectors $\vec{U}_1$ and $\vec{U}_2$, as is shown in Fig. 8, and considering a enough short switching period, it is:

$$\vec{U}^* \cdot T_z = \vec{U}_1 \cdot T_1 + \vec{U}_2 \cdot T_2$$

$$\vec{U}^* = \vec{U}_1 \frac{T_1}{T_z} + \vec{U}_2 \frac{T_2}{T_z} \tag{19}$$

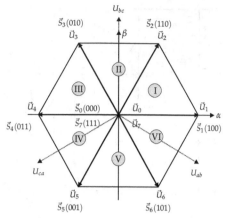

Figure 6. Switching and voltage vectors

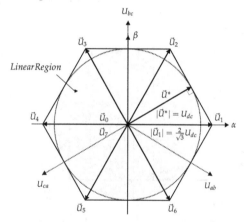

Figure 7. Linear region to work without overmodulation

The times T_1 and T_2 are calculated using trigonometric projections as is shown in Fig. 8, it is:

$$T_1 = \frac{|\vec{U}^*|}{|\vec{U}_1|} \cdot T_z \frac{\sin(\frac{\pi}{3} - \phi)}{\sin(\frac{2\pi}{3})} \tag{20}$$

$$T_2 = \frac{|\vec{U}^*|}{|\vec{U}_2|} \cdot T_z \frac{\sin(\phi)}{\sin(\frac{2\pi}{3})} \tag{21}$$

Where T_1 and T_2 are the times of application of the active vectors in a switching period, T_Z is the switching period and ϕ is the angle between the reference voltage vector and the adjacent active vector ($\vec{U}_1$). If the sum of times T_1 and T_2 is minor of the switching period, the rest of the time is apply the zero vectors, it is:

$$T_0 = T_7 = T_Z - T_1 - T_2 \tag{22}$$

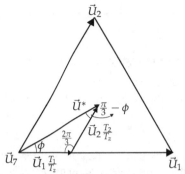

Figure 8. Voltage vector $\vec{U}^*$ and its components in sector I

Where T_0 and T_7 are the times of applications of zero vectors in a switching period. Once calculated the times of applications of each adjacent voltage vectors the next step is to follow a specific switching sequence for the symmetrical space vector modulation technique, this one depends if the reference vector is localized in an even or odd sector, e.g. in Fig. 9 is observed the optimum switching sequence and the pulse pattern for odd sector ($\vec{S}_0, \vec{S}_1, \vec{S}_2$ and $\vec{S}_7$), however for even sector the switching sequence is contrary to the case for odd sector as is shown in Fig. 10.

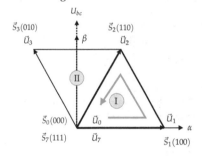
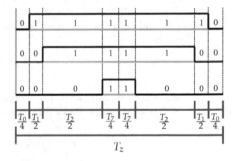

Figure 9. Switching sequence for odd sector.

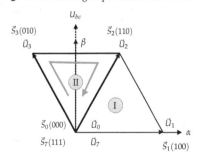
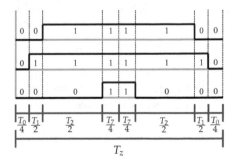

Figure 10. Switching sequence for even sector.

The details about the implementation of SVM algorithm in MATLAB/Simulink are presented in [3] in page 97.

3.3. Electromagnetic torque and stator flux estimation

The Figure 3 shows that the electromagnetic torque and the stator flux estimation depends of the stator voltage and the stator current space vectors, therefore:

$$\vec{\psi}_s = \int (\vec{u}_s - R_s \cdot \vec{i}_s)dt \tag{23}$$

The problem in this type of estimation is when in low speeds the back electromotive force (emf) depends strongly of the stator resistance, to resolve this problem is used the current model to improve the flux estimation as in [17]. The rotor flux represented in the rotor flux reference frame is:

$$\vec{\psi}_{rdq} = \frac{L_m}{1 + sT_r}\vec{i}_{sdq} - j\frac{(\omega_{\psi_r} - \omega_r)T_r}{1 + sT_r}\vec{\psi}_{rdq} \tag{24}$$

Where $T_r = L_r/R_r$ is the rotor time constant. In this reference frame $\psi_{rq} = 0$ and substituting this expression in the equation (24), it is:

$$\psi_{rd} = \frac{L_m}{1 + sT_r}i_{sd} \tag{25}$$

In the current model the stator flux is represented as:

$$\vec{\psi}_s^i = \frac{L_m}{L_r}\vec{\psi}_r^i + \frac{L_sL_r - L_m^2}{L_r}\vec{i}_s \tag{26}$$

Where $\vec{\psi}_r^i$ is the rotor flux estimated in the equation (25). The voltage model is based in the equation (1) and from there the stator flux in the stationary reference frame is:

$$\vec{\psi}_s = \frac{1}{s}(\vec{v}_s - R_s\vec{i}_s - \vec{U}_{comp}) \tag{27}$$

With the aim to correct the errors associated with the pure integration and the stator resistance variations with temperature, the voltage model is adapted through the PI controller.

$$\vec{U}_{comp} = (K_p + K_i\frac{1}{s})(\vec{\psi}_s - \vec{\psi}_s^i) \tag{28}$$

The K_p and K_i coefficients are calculated with the recommendation proposed in [17]. The rotor flux $\vec{\psi}_r$ in the stationary reference frame is calculated as:

$$\vec{\psi}_r = \frac{L_r}{L_m}\vec{\psi}_s - \frac{L_sL_r - L_m^2}{L_m}\vec{i}_s \tag{29}$$

The estimator scheme shown in the Figure 11 works with a good performance in the wide range of speeds.

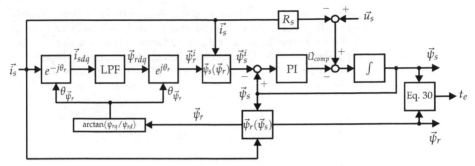

Figure 11. Stator and rotor flux estimator, and electromagnetic torque estimator.

Where LPF means low pass filter. In the other hand, when the equations (27) and (29) are replaced in (5) it is estimated the electromagnetic torque.

$$t_e = \frac{3}{2} P \frac{L_m}{L_r L_s \sigma} \vec{\psi}_r \times \vec{\psi}_s \qquad (30)$$

4. Design of Takagi-Sugeno fuzzy controller

The Takagi-Sugeno Fuzzy controller takes as inputs the stator flux error E_{ψ_s} and the electromagnetic torque error E_τ, and as outputs the quadrature components of the stator voltage vector, represented in the stator flux reference frame. The first output (u_{ds}^*) takes a linear combination of the inputs as a consequent part of the rules, similarly, the second output (u_{qs}^*) takes the similar linear combination used in the first output but with the coefficients interchanged how is shown in the Figure 12.

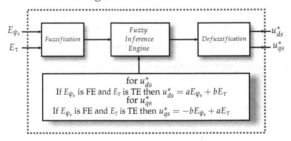

Figure 12. The structure of a fuzzy logic controller.

4.1. Membership functions

The Membership Functions (MF) for T-S fuzzy controller are shown in Figure 13 and in Figure 14, for the stator flux error and the electromagnetic torque error, respectively. These MF's shape and parameters was found through trial and error method with multiple simulations and with the knowing of the induction motor response for every test. This method is know as subjective approach [22].

The universe of discourse for the stator flux error input is defined in the closed interval [-0.5, 0.5]. The extreme MFs have trapezoidal shapes but the middle one takes triangular shape as is shown in Figure 13. However, the universe of discourse for electromagnetic torque error input is defined in the closed interval [-20, 20] but with the objective to see the shape of the MFs only is shown the interval [-5, 5] in Figure 14, the shapes of these MF are similar to the first input. For both inputs the linguistic labels N, Ze and P means Negative, Zero and Positive, respectively.

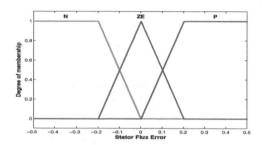

Figure 13. Membership function for stator flux error input (E_{ψ_s})

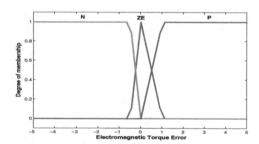

Figure 14. Membership function for electromagnetic torque error input (E_τ)

4.2. The fuzzy rule base

The direct component of the stator voltage u_{ds}^* is determined by the rules of the form:

$$R_x : \text{ if } E_{\psi_s} \text{ is } FE \text{ and } E_\tau \text{ is } TE \text{ then } u_{ds}^* = aE_{\psi_s} + bE_\tau$$

However, the quadrature component of the stator voltage u_{qs}^* is determined by the rules of the form:

$$R_y : \text{ if } E_{\psi_s} \text{ is } FE \text{ and } E_\tau \text{ is } TE \text{ then } u_{qs}^* = -bE_{\psi_s} + aE_\tau$$

Where $FE = TE = \{N, ZE, P\}$ are the fuzzy sets of the inputs and, **a** and **b** are coefficients of the first-order polynomial function typically present in the consequent part of the firs-order Takagi-Sugeno fuzzy controllers.

For instance, when the consequent function of the rule R_i is a real number, the consequent function is a zero-order polynomial and we have a zero-order controller. If the consequent function is a linear we have first-order controller [21].

$$R_i : \text{ if } X \text{ is } A_i \text{ and } Y \text{ is } B_i \text{ then } z = f_i(X, Y)$$

The rule base to calculate u_{ds}^* and u_{qs}^* is shown in Table 2. The **product** is the conjunction operator and the weighted average (**wtaver**) is the defuzzification method used to set the controller in the MATLAB fuzzy editor.

E_{-s} / E_o	N	ZE	P
N	$u_{ds}^* = aF_e + bT_e$ $u_{qs}^* = -bF_e + aT_e$	$u_{ds}^* = aF_e + bT_e$ $u_{qs}^* = -bF_e + aT_e$	$u_{ds}^* = aF_e + bT_e$ $u_{qs}^* = -bF_e + aT_e$
ZE	$u_{ds}^* = aF_e + bT_e$ $u_{qs}^* = -bF_e + aT_e$	$u_{ds}^* = aF_e + bT_e$ $u_{qs}^* = -bF_e + aT_e$	$u_{ds}^* = aF_e + bT_e$ $u_{qs}^* = -bF_e + aT_e$
P	$u_{ds}^* = aF_e + bT_e$ $u_{qs}^* = -bF_e + aT_e$	$u_{ds}^* = aF_e + bT_e$ $u_{qs}^* = -bF_e + aT_e$	$u_{ds}^* = aF_e + bT_e$ $u_{qs}^* = -bF_e + aT_e$

Table 2. Fuzzy rules for computation of u_{ds}^* and u_{qs}^*

5. Simulation results

The simulations were performed using MATLAB simulation package which include Simulink block sets and fuzzy logic toolbox. The switching frequency of PWM inverter was set to be $10kHz$, the stator reference flux considered was 0.47 Wb and the coefficients considered were $a = 90$ and $b = 2$. In order to investigate the effectiveness of the proposed control system and in order to check the closed-loop stability of the complete system, we performed several tests.

We used different dynamic operating conditions such as: step change in the motor load (from 0 to 1.0 pu) at fifty percent of rated speed, no-load sudden change in the speed reference (from 0.5 pu to -0.5 pu), and the application of an arbitrary load torque profile at fifty percent of rated speed. The motor parameters are given in Table 3.

Rated voltage (V)	220/60Hz
Rated Power (HP)	3
Rated Torque (Nm)	11.9
Rated Speed (rad/s)	179
$R_s, R_r (\Omega)$	0.435, 0.816
L_{ls}, L_{lr} (H)	0.002, 0.002
L_m (H)	0.0693
$J (K_g m^2)$	0.089
P (pole pairs)	2

Table 3. Induction Motor Parameters [16]

The Figure 15 illustrates the torque response of the DTC-SVM scheme with T-S fuzzy controller when the step change in the motor load is apply. The electromagnetic torque tracked the reference torque and in this test is obtained the following good performance measures: rise time $t_r = 1.1ms$, settling time $t_s = 2.2ms$ and torque ripple $ripple = 2.93\%$. Also is observed that the behavior of the stator current is sinusoidal.

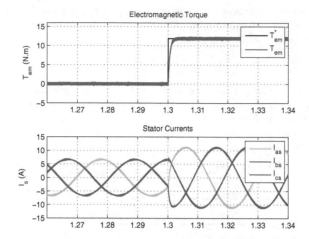

Figure 15. Electromagnetic torque and stator current response for step change in the motor load at fifty percent of rated speed

The Figure 16 presents the results when an arbitrary torque profile is imposed to DTC-SVM scheme with T-S fuzzy controller. In the first sub-figure the electromagnetic torque tracked the reference torque as expected, and in the next one the sinusoidal waveforms of the stator currents is shown. The Figure 17 shows space of the quadrature components of the stator flux and it shows the circular behavior of the stator flux when the torque profile is applied, and in consequence the proposed controller maintain the stator flux constant.

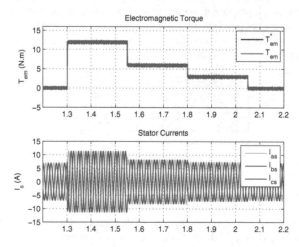

Figure 16. Electromagnetic torque and stator current response when is apply the load torque profile at fifty percent of rated speed

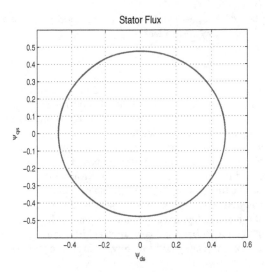

Figure 17. Space of the stator flux quadrature components.

The Figure 18 shows the behavior of the rotor angular speed ω_r, the electromagnetic torque and the phase a stator current waveform when a step change in the reference speed from 0.5 pu to -0.5 pu is imposed, with no-load. The torque was limited in 1.5 times the rated torque how it was projected and the sinusoidal waveforms of the stator current shown that this control technique allowed also a good current control because it is inherent to the algorithm control

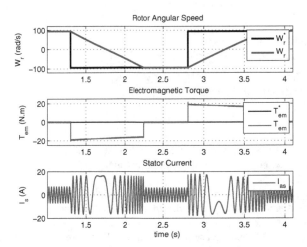

Figure 18. Rotor angular speed, electromagnetic torque and phase a stator current when was apply the no-load sudden change in the speed reference at fifty percent of rated speed

proposed in this chapter. All the test results showed the good performance of the proposed DTC-SVM scheme with T-S fuzzy controller.

6. Conclusion

This chapter presents the DTC-SVM scheme with T-S fuzzy controller for the three-phase IM. The conventional DTC-SVM scheme takes two PI controllers to generate the reference stator voltage vector. To improve the drawback of this conventional DTC-SVM scheme is proposed the Takagi-Sugeno fuzzy controller to substitute both PI controllers. The proposed controller calculates the quadrature components of the reference stator voltage vector in the stator flux reference frame. The rule base for the proposed controller is defined in function of the stator flux error and the electromagnetic torque error using trapezoidal and triangular membership functions. The direct component of the stator voltage takes a linear combination of its inputs as a consequent part of the rules, however, the quadrature component of the stator voltage takes the similar linear combination used in the first output but with the coefficients interchanged, not to be necessary another different coefficients values for this output. Constant switching frequency and low torque ripple are obtained using space vector modulation technique.

Simulations at different operating conditions have been carried out. The simulation results verify that the proposed DTC-SVM scheme with T-S fuzzy controller achieved good performance measures such as rise time, settling time and torque ripple as expected, It shown the fast torque response and low torque ripple in a wide range of operating conditions such as step change in the motor load, no-load sudden change in the speed reference, and the application of an arbitrary load torque profile. These results validate the proposed scheme.

Author details

José Luis Azcue and Ernesto Ruppert
University of Campinas (UNICAMP), Brazil

Alfeu J. Sguarezi Filho
CECS/UFABC, Santo André - SP, Brazil

7. References

[1] Abu-Rub, H., Guzinski, J., Krzeminski, Z. & Toliyat, H. [2004]. Advanced control of induction motor based on load angle estimation, *Industrial Electronics, IEEE Transactions on* 51(1): 5 – 14.

[2] Azcue P., J. L. [2009]. Modelamento e simulação do motor de indução trifásico. URL: *http://www.mathworks.fr/matlabcentral/fileexchange/24403-modelamento-e-simulação-do-motor-de-indução-trifásico*

[3] Azcue P., J. L. [2010]. *Three-phase induction motor direct torque control using self-tuning pi-type type fuzzy controller.*, Master's thesis, University of Campinas (UNICAMP). URL: *http://cutter.unicamp.br/document/?code=000777279*

The Takagi-Sugeno Fuzzy Controller Based Direct Torque Control with Space Vector Modulation for
Three-Phase Induction Motor

91

[4] Azcue P., J. & Ruppert, E. [2010]. Three-phase induction motor dtc-svm scheme with self-tuning pi-type fuzzy controller, *Fuzzy Systems and Knowledge Discovery (FSKD), 2010 Seventh International Conference on*, Vol. 2, pp. 757 –762.

[5] Bertoluzzo, M., Buja, G. & Menis, R. [2007]. A direct torque control scheme for induction motor drives using the current model flux estimation, *Diagnostics for Electric Machines, Power Electronics and Drives, 2007. SDEMPED 2007. IEEE International Symposium on* pp. 185 –190.

[6] Buja, G. & Kazmierkowski, M. [2004]. Direct torque control of pwm inverter-fed ac motors - a survey, *Industrial Electronics, IEEE Transactions on* 51(4): 744–757.

[7] Cao, S., Liu, G. & Cai, B. [2009]. Direct torque control of induction motors based on double-fuzzy space vector modulation technology, *Information Engineering and Computer Science, 2009. ICIECS 2009. International Conference on*, pp. 1 –4.

[8] Chen, L., Fang, K.-L. & Hu, Z.-F. [2005]. A scheme of fuzzy direct torque control for induction machine, *Machine Learning and Cybernetics, 2005. Proceedings of 2005 International Conference on*, Vol. 2, pp. 803 –807 Vol. 2.

[9] Depenbrock, M. [1988]. Direct self-control (dsc) of inverter-fed induction machine, *Power Electronics, IEEE Transactions on* 3(4): 420 –429.

[10] Ding, X., Liu, Q., Ma, X., He, X. & Hu, Q. [2007]. The fuzzy direct torque control of induction motor based on space vector modulation, *Natural Computation, 2007. ICNC 2007. Third International Conference on*, Vol. 4, pp. 260 –264.

[11] Habetler, T., Profumo, F., Pastorelli, M. & Tolbert, L. [1992]. Direct torque control of induction machines using space vector modulation, *Industry Applications, IEEE Transactions on* 28(5): 1045 –1053.

[12] Holtz, J. [1992]. Pulsewidth modulation-a survey, *Industrial Electronics, IEEE Transactions on* 39(5): 410 –420.

[13] Jiang, Z., Hu, S. & Cao, W. [2008]. A new fuzzy logic torque control scheme based on vector control and direct torque control for induction machine, *Innovative Computing Information and Control, 2008. ICICIC '08. 3rd International Conference on*, p. 500.

[14] Kang, J.-K. & Sul, S.-K. [1999]. New direct torque control of induction motor for minimum torque ripple and constant switching frequency, *Industry Applications, IEEE Transactions on* 35(5): 1076 –1082.

[15] Koutsogiannis, Z., Adamidis, G. & Fyntanakis, A. [2007]. Direct torque control using space vector modulation and dynamic performance of the drive, via a fuzzy logic controller for speed regulation, *Power Electronics and Applications, 2007 European Conference on*, pp. 1 –10.

[16] Krause, P. C., Wasynczuk, O. & Sudhoff, S. D. [2002]. *Analysis of Electric Machinery and Drive Systems*, IEEE Press.

[17] Lascu, C., Boldea, I. & Blaabjerg, F. [2000]. A modified direct torque control for induction motor sensorless drive, *Industry Applications, IEEE Transactions on* 36(1): 122–130.

[18] Lin, G. & Xu, Z. [2010]. Direct torque control of induction motor based on fuzzy logic, *Computer Engineering and Technology (ICCET), 2010 2nd International Conference on*, Vol. 4, pp. V4–651 –V4–654.

[19] Pan, Y. & Zhang, Y. [2009]. Research on direct torque control of induction motor based on dual-fuzzy space vector modulation technology, *Fuzzy Systems and Knowledge Discovery, 2009. FSKD '09. Sixth International Conference on*, Vol. 6, pp. 383 –388.

[20] Park, Y.-M., Moon, U.-C. & Lee, K. [1995]. A self-organizing fuzzy logic controller for dynamic systems using a fuzzy auto-regressive moving average (farma) model, *Fuzzy Systems, IEEE Transactions on* 3(1): 75 –82.

[21] Pedrycz, W. & Gomide, F. [2007]. *Fuzzy Systems Engineering Toward Human-Centric Computing*, Wiley-IEEE Press.

[22] Reznik, L. [1997]. *Fuzzy Controllers Handbook: How to Design Them, How They Work*, Newnes.

[23] Takahashi, I. & Noguchi, T. [1986]. A new quick-response and high-efficiency control strategy of an induction motor, *Industry Applications, IEEE Transactions on* IA-22(5): 820 –827.

[24] van der Broeck, H., Skudelny, H. & Stanke, G. [1988]. Analysis and realization of a pulsewidth modulator based on voltage space vectors, *Industry Applications, IEEE Transactions on* 24(1): 142–150.

[25] Vas, P. [1998]. *Sensorless vector and Direct Torque Control*, Oxford University Press.

[26] Viola, J., Restrepo, J., Guzman, V. & Gimenez, M. [2006]. Direct torque control of induction motors using a fuzzy inference system for reduced ripple torque and current limitation, *Power Electronics and Motion Control Conference, 2006. EPE-PEMC 2006. 12th International*, pp. 1161 –1166.

[27] Zhou, K. & Wang, D. [2002]. Relationship between space-vector modulation and three-phase carrier-based PWM: a comprehensive analysis [three-phase inverters], *Industrial Electronics, IEEE Transactions on* 49(1): 186–196.

Chapter 5

Evaluation of an Energy Loss-Minimization Algorithm for EVs Based on Induction Motor

Pedro Melo, Ricardo de Castro and Rui Esteves Araújo

Additional information is available at the end of the chapter

1. Introduction

This work addresses the problem of optimal selection of the flux level in induction motors used in electric vehicles (EVs). The basic function of a fully electric powertrain controller is to generate electric torque (force) which is required at any time by the driver. But, it is well-known that the flux level used in a controller for induction motors offers an extra degree of freedom that can be used to maximise energy efficiency. The induction motor is an efficient motor when working close to its rated operating point (Zeraoulia, Benbouzid et al. 2006). However, at light loads the efficiency is greatly reduced when magnetization flux is maintained at nominal value. In induction motor drives for EVs, where real operation conditions are significantly different from rated conditions, the energy saving control is crucial for improving the running distance per charge.

Due to the widespread use of induction motors, its efficiency optimization gave rise to a large number of research publications (Bazzi & Krein 2010). Algorithms for real-time implementation of loss-minimization methods are vital for designing intelligent and optimized EV controllers. Standard methods for induction motor control, including field-oriented control (FOC) or direct torque control (DTC), can be improved in efficiency by using loss minimization control. Basically, there are three different methods to improve the efficiency in induction motors: i) loss model based methods (which is considered in this work), ii) power measure based methods, also known as search controllers; and iii) hybrid controllers that combines the first two methods. The main goal of the present work is to investigate the potential benefits of loss minimization algorithms in EVs powered by induction motors. Accordingly, a detailed simulation case study will be provided which will show that, depending on the type of driving cycle, energy savings up to 12.5% can be achieved. The chapter is organized as follows: Section 2 reviews the basic concepts of rotor field oriented control (FOC). Section 3 introduces the loss minimization method based on a standard mathematical model of the induction motor and gives the value of the flux level which maximizes the energy efficiency at given torque subject to voltage and currents limits. In Section 4 the developed EV non-causal simulation model (motor-to-wheel) is presented, while Section 5 includes the simulation results and its analysis for a set of standard driving cycles. Finally, Section 6 contains the main conclusions and some reference to future work.

2. Rotor FOC

In this study, a model based approach was selected for minimizing the induction motor losses (Lim & Nam 2004). This Loss Minimization Algorithm (LMA) was developed in d-q coordinates, considering an equivalent motor model in the synchronous reference frame, as described in section 3. In addition, as we will discuss in a later section, the induction motor controller is also based on rotor FOC. These reasons justify a brief review on induction motor rotor FOC. Figure 1 represents the basic concept of rotor FOC (based on Krishnan, 2001).

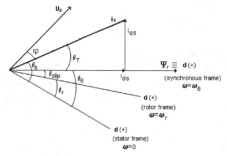

Figure 1. Rotor FOC principle for induction motors

Recall that, in the synchronous frame (ω_e), the rotor magnetic flux (Ψ_{rd}) is aligned with d axis, thus $\Psi_{rd} = \Psi_r$, $\Psi_{rq} = 0$. In the same reference frame, the stator current component i_{ds} is aligned with the rotor magnetic flux, controlling its value. On the other hand, i_{qs} (shifted $\pi/2$ electrical rad from i_{ds}) controls the motor electromagnetic torque:

$$\Psi_r = L_m i_{ds} \left(\text{steady} - \text{state}\right) \tag{1}$$

$$T_t\left(t\right) = K_t \Psi_r i_{qs} \tag{2}$$

From figure 1, it may be seen that i_{ds} and i_{qs} are, respectively, the d and q components of the space vector $\mathbf{i_s}$ in the synchronous reference frame. This way, from the control philosophy perspective, i_{ds} and i_{qs} regulation is implemented in this reference frame; however, from the control hardware perspective, i_{ds} and i_{qs} must be considered in stator reference phase-coordinates (i_a, i_b, i_c). To do that, it is mandatory to obtain i_{ds} and i_{dq} in the static d-q reference, which requires the information about $\theta_s = \theta_e + \theta_T$. The determination of θ_e is the main issue, since $\theta_T = arctg(i_{qs}/i_{ds})$; θ_e calculation can be accomplished through θ_{slip} and θ_r (see figure 1) – indirect FOC.

Since
$$\omega_{\text{slip}} = \omega_e - \omega_r = \frac{L_m}{L_r / R_r} \frac{i_{qs}}{\Psi_r} \tag{3}$$

θ_{slip} is given by:

$$\theta_{\text{slip}}\left(t\right) = \theta_{\text{slip}}\left(t_0\right) + \int_{t_0}^{t} \omega_{\text{slip}} \, dt \tag{4}$$

Knowing the instantaneous rotor speed ω_r, one have:

$$\theta_r\left(t\right) = \theta_r\left(t_0\right) + \int_{t_0}^{t} \omega_r dt \tag{5}$$

From figure 1:

$$\theta_e\left(t\right) = \theta_{\text{slip}}\left(t\right) + \theta_r\left(t\right) \tag{6}$$

3. Loss minimization by selecting flux references

The loss-minimization scheme demands the decrease or increase of the flux level depending on the torque. This means that the minimization algorithm selects the flux reference through the minimization of the copper and core losses while ensuring the desired torque requested by the driver. Different techniques for loss minimization in induction motor are presented in the literature (Bazzi & Krein 2010). Recently, (Lim & Nam 2004) proposed a LMA that features a major difference from previous works by taking into consideration the leakage inductance and the practical constrains on voltage and current in the high-speed region, which play a great role in EVs applications. This is an important difference from other works, like (Garcia et al., 1994), (Kioskeridis & Margaris, 1996), (Fernandez-Bernal et al., 2000), where leakage inductance are not considered (although similar motor loss models are included), leading to considerable result differences in the high-speed region. In addition, our work considers the optimization of both positive and negative torque generation with bounded constraints on both current and voltage.

3.1. The LMA method

The implemented method is based on the conventional induction motor model where the iron losses are represented by an equivalent resistance (R_m) modelling the iron losses, placed in parallel with the magnetizing inductance (L_m). A simplification is then considered, allowing a partial decoupling between R_m and L_m: the iron losses are represented by separated circuits with dependent voltage sources (V_{dm}^e and V_{qm}^e). Figure 2 shows the complete equivalent model in the synchronous reference frame.

Considering steady state analysis with low slip values (s) – rotor iron losses may be neglected –, the total motor losses (copper and iron ones) are given by (Lim & Nam 2004):

$$P_{loss} = R_d\left(\omega_e\right)i_{ds}^{e\,2} + R_q\left(\omega_e\right)i_{qs}^{e\,2} \tag{7}$$

$$R_d\left(\omega_e\right) = R_s + \frac{\omega_e^2 L_m^2}{R_m} \tag{8}$$

$$R_q\left(\omega_e\right) = R_s + \frac{R_r L_m^2}{L_r^2} + \frac{\omega_e^2 L_m^2 L_{lr}^2}{R_m L_r^2} \tag{9}$$

Where:

i_{ds}^e : d-axis stator current in the synchronous reference frame;

i_{qs}^e : q-axis stator current in the synchronous reference frame;

ω_e : electrical angular frequency;

- Rs; Rr: stator and rotor resistances (respectively);
- Rm: equivalent stator iron losses resistance;
- Lr; Llr: rotor total inductance and rotor leakage inductance (respectively);
- Lm: magnetizing inductance.

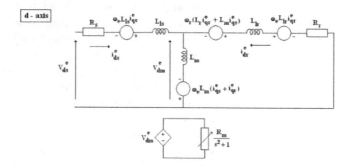

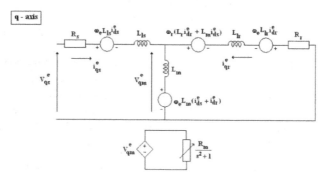

Figure 2. Simplified motor equivalent model (Lim & Nam 2004)

Note that $R_d(\omega_e)$ and $R_q(\omega_e)$ are the direct (d) and quadrature (q) components of the equivalent resistors representing the total losses. Voltage and current constraints (mentioned before) are defined by (neglecting stator resistor drop):

$$(\omega_e L_s i^e_{ds})^2 + (\omega_e \sigma L_s i^e_{qs})^2 \leq V^2_{max} \tag{10}$$

$$i^e_{ds}{}^2 + i^e_{qs}{}^2 \leq I^2_{max} \tag{11}$$

Where:

$$\sigma = 1 - L^2_m / (L_s L_r) \tag{12}$$

σ: induction machine leakage coeficient; L_s: stator total inductance;

V_{max}; I_{max}: motor (or inverter) voltage and current limits, respectively;

An important observation is that voltage constraint depends on the considered ω_e.

The LMA's goal is to achieve the optimal flux level that minimizes the motor total losses under voltage and current constraints. The motor rated flux level must also be taken into consideration, in order to avoid magnetic saturation. Moreover, the torque developed by the motor cannot be compromised by the LMA implementation. From the mathematical point of view, the LMA algorithm consists in:

$$\min P_{loss}(i^e_{ds}, i^e_{qs})$$

$$s.t. : (10), (11)$$

$$i^e_{ds} \leq I_{dn} \tag{13}$$

$$T_e = K_t i^e_{ds} i^e_{qs}$$

Where:

I_{dn} : rated d-axis stator current

T_e : electromagnetic torque (steady-state), considering rotor FOC;

$$K_t = \frac{3}{2} p \frac{L^2_m}{L_r} \tag{14}$$

[p: pairs of magnetic poles]

3.1.1. Unconstrained optimization

In the (i_{ds}^e, i_{qs}^e) domain, the optimal flux solution for the region inside the inequality restrictions is achieved through Lagrange multipliers method, since only one restriction is active – the torque one

For one restriction only, the general problem is formulated as follows:

$$\nabla L(i_{ds}^{\ e}, i_{qs}^{\ e}, \lambda) = 0 \tag{15}$$

with:

$$L(i_{ds}^{\ e}, i_{qs}^{\ e}, \lambda) = P_{loss}(i_{ds}^{\ e}, i_{qs}^{\ e}) + \lambda(T_e - K_t i_{ds}^{\ e} i_{qs}^{\ e}) \tag{16}$$

where L(i_{ds}^e, i_{qs}^e, λ) is the lagrangian associated to the problem, λ is the Lagrange multiplier, P_{loss}(i_{ds}^e, i_{qs}^e) is the cost function and T_e-$K_t i_{ds}^e i_{qs}^e$ is the restriction. Applying first-order optimal condition (15) gives the following equation system:

$$\frac{\partial L}{\partial i_{sq}^{\ e}} = 0 \qquad \frac{\partial L}{\partial i_{sq}^{\ e}} = 0 \qquad \frac{\partial L}{\partial \lambda} = 0 \tag{17}$$

yielding

$$i_{ds}^e = \left(\frac{T_e^2 \, R_q(\omega_e)}{K_t^2 \, R_d(\omega_e)} \right)^{1/4} \ ; \ i_{qs}^e = \left(\frac{T_e^2 \, R_d(\omega_e)}{K_t^2 \, R_q(\omega_e)} \right)^{1/4} \tag{18}$$

3.1.2. Constrained optimization

Previously, all the inequalities were considered inactive. In order to obtain the optimal solutions in each restriction boundary, the Lagrange multipliers method is applied for each inequality constraint activation (i.e. only "=" operator is valid), together with the torque one. This way, three non linear algebraic equation systems are defined for the inequality constraints. The optimal i_{ds}^e is given by these systems solutions, since it refers to regions on the border lines of the inequality restrictions.

Table 1 presents the solutions, in (i_{ds}^e, i_{qs}^e) plane, for interior points (zone 0) and for inequality restriction borders (zones 1, 2 and 3).

The voltage and current limits (V_{max}, I_{max} and I_{dn}) lead naturally to three regions of operation referred to as constant torque (low-speed), constant power (midrange speed) and constant power-speed (high-speed), as defined in (Novotny & Lipo, 1996). The transition between constant torque region and power region is characterized by the rated speed (ω_n), which is defined by the interception of inequality restrictions border lines:

$$(\omega_n L_s i_{ds}^e)^2 + (\omega_n \sigma L_s i_{qs}^e)^2 = V_{max}^2 \tag{19}$$

$$i_{ds}^{e\ 2} + i_{qs}^{e\ 2} = I_{max}^2 \tag{20}$$

$$i_{ds}^e = I_{dn} \tag{21}$$

Zone	Name	Active Constraints	Solution
0	LMA Operation in Interior Points	$T_e = K_t i_{ds}^e i_{qs}^e$	(18)
1	Max Torque Limit	$T_e = K_t i_{ds}^e i_{qs}^e$ $i_{ds}^e = I_{dn}$	$i_{ds}^e = I_{dn} ; \; i_{qs}^e = \dfrac{T_e}{K_t I_{dn}}$
2	Max. Current Limit	$T_e = K_t i_{ds}^e i_{qs}^e$ $i_{ds}^{e\,2} + i_{qs}^{e\,2} = I_{max}^2$	$i_{ds}^e = \left(\dfrac{I_{max}^2 - (I_{max}^4 - 4T_e^2 / K_t^2)^{1/2}}{2} \right)^{1/2}$ $i_{qs}^e = \left(\dfrac{I_{max}^2 + (I_{max}^4 - 4T_e^2 / K_t^2)^{1/2}}{2} \right)^{1/2}$
3	Max. Voltage Limit	$T_e = K_t i_{ds}^e i_{qs}^e$ $(\omega_e L_s i_{ds}^e)^2$ $+ (\omega_e \sigma L_s i_{qs}^e)^2 = V_{max}^2$	$i_{ds}^e = \left(\dfrac{V_{max}^2 + (V_{max}^4 - 4\omega_e^4 \sigma^2 L_s^4 T_e^2 / K_t^2)^{1/2}}{2(\omega_e L_s)^2} \right)^{1/2}$ $i_{qs}^e = \left(\dfrac{V_{max}^2 - (V_{max}^4 - 4\omega_e^4 \sigma^2 L_s^4 T_e^2 / K_t^2)^{1/2}}{2(\omega_e L_s)^2} \right)^{1/2}$

Table 1. LMA optimized solutions

The calculated result is:

$$\omega_n = \frac{V_{max}}{L_s} \frac{1}{[I_{dn}^2 + \sigma^2 (I_{max}^2 - I_{dn}^2)]^{1/2}} \tag{22}$$

ω_c is the boundary speed between constant power and power-speed ($P_{mec}*\omega_e$=constant) regions:

$$\omega_c = \frac{V_{max}}{I_{max} L_s} \left(\frac{\sigma^2 + 1}{2\sigma^2} \right)^{1/2} \tag{23}$$

For region 1, the maximum torque is limited by I_{dn} and I_{max}:

$$T_{m1} = K_t I_{dn} (I_{max}^2 - I_{dn}^2)^{1/2} \tag{24}$$

The maximum torque in region 2 is limited by V_{max} and I_{max}:

$$T_{m2} = K_t \frac{[(V_{max}/(\omega_e L_s))^2 - I_{max}^2 \sigma^2]^{1/2} [I_{max}^2 - (V_{max}/(\omega_e L_s))^2]^{1/2}}{1 - \sigma^2} \tag{25}$$

In region 3, the maximum torque is limited by V_{max}, but the current is smaller than I_{max}. So, the current limit does not interfere with T_{m3}:

$$T_{m3} = K_t \left(\frac{V_{max}}{\omega_e L_s} \right)^2 \frac{1}{2\sigma} \tag{26}$$

3.1.3. Optimal I_ds generation

For the zone in the (i_{ds}^e, i_{qs}^e) plane limited by restrictions (10), (11) and $i_{ds}^e \leq I_{dn}$, optimal result (18) is valid, meaning that:

$$i_{ds}^e = \left(\frac{R_q(\omega_e)}{R_d(\omega_e)} \right)^{1/2} i_{qs}^e \tag{27}$$

In the border lines of those restrictions, the previous relation can not be considered. So, for region 1, (18) is applied if:

$$i_{qs}^e \leq \left(\frac{R_d(\omega_e)}{R_q(\omega_e)} \right)^{1/2} * I_{dn} \tag{28}$$

The i_{qs}^e upper limit in (28) defines T_{p1}(see Figure 3):

$$T_{p1} = K_t \left(\frac{Rd(\omega_e)}{Rq(\omega_e)} \right)^{1/2} I_{dn}^2 \tag{29}$$

Of course, for: $\left(\dfrac{Rd(\omega_e)}{Rq(\omega_e)} \right)^{1/2} * I_{dn} < i_{qs}^e \leq (I_{max}^2 - I_{dn}^2)^{1/2} \rightarrow i_{ds}^e = I_{dn}$ (30)

For region 2, (18) can be considered, until the voltage limit (V_{max}) is achieved:

$$i_{qs}^e \leq \frac{V_{max} / (\omega_e L_s \sigma)}{[\sigma^{-2} + R_d(\omega_e)/R_q(\omega_e)]^{1/2}} \left(\frac{Rd(\omega_e)}{Rq(\omega_e)} \right)^{1/2} \tag{31}$$

This way, T_{p2} is given by the following expression:

$$T_{p2} = K_t \frac{V_{max}^2 [R_d(\omega_e) * R_d(\omega_e)]^{1/2}}{[\sigma^2 R_d(\omega_e) + R_q(\omega_e)](\omega_e L_s)^2} \tag{32}$$

Above this limit, i_{ds}^e (and i_{qs}^e) is given by zone 3 solution (table 1).

As stated before, only the voltage limit must be considered for region 3, which means that $T_{p3} = T_{p2}$. Of course, for this region one must consider $\omega_e > \omega_c$.

Figure 3 presents the paths for I_{ds}^* generation in i_{ds}; i_{qs} coordinates (origin-T_p-T_m), considering the three described operation regions. Quadrants I and II are represented, in order to consider both motor and braking modes (optimal i_{ds} paths for quadrant II are symmetric to quadrant I paths).

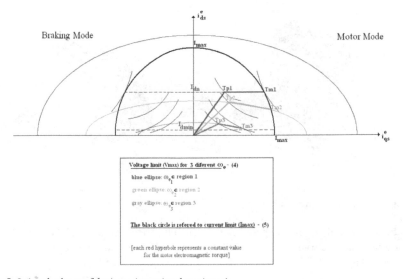

Figure 3. I_{ds}^* paths for ω_{e1} (blue), ω_{e2} (green) and ω_{e3} (gray)

It is clear the linear evolution in the three regions (given by (27)), while in region 3, only voltage limit must be considered, since I_{max} is not reached. After that, in region 1, I_{dn} imposes the optimal path. In region 2 both current and voltage limits (i.e. I_{max} and V_{max}) restrict I_{ds} optimal path, while in region 3, I_{max} most probably is not reached.

3.2. Optimal I$_{ds}$ generation for the simulated induction motor

In order to get some insight on LMA main features, a first set of results is presented in figures 4-6, based on an induction motor, with the following parameters:

[Rs; Rr] (Ω) [0,399; 0,3538]
[Ls; Lr] (H) [59,3; 60,4]*10^{-3}
[ls; lr] (H) [2,7; 3,8]*10^{-3}
Lm (H) 56,6*10^{-3}
Rm (Ω) 350
J(kg m^2) 0,089

Table 2. Induction Motor Parameters (9 kW; 60 Hz; 4 poles; 1750 rpm)

Figure 4 represents the optimal I_d generation for conventional approach, i.e. constant flux +field weakening (CF+FW), and the LMA approach.

Inspecting these results one can find that the LMA influence on I_d* is mostly visible for low torques (T<20 N.m). It is interesting to note that in the high speed zone (>2000 rpm), LMA and conventional flux regulation tend to present closer I_d* values, as the speed increases. Also, for high torque values (above 30 N.m) both approaches have similar performances.

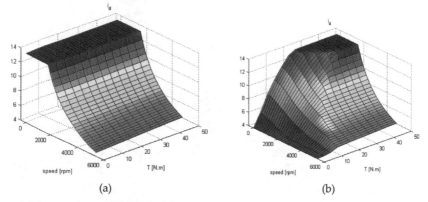

(a)   (b)

Figure 4. $I_{ds}*$ generations: a) CF+FW; b) LMA

From the above analysis, it is expectable that the differences in the generation of I_d* lead to different efficiencies curves of the induction motor, which is, indeed, observed in the maps illustrated in Figure 5.

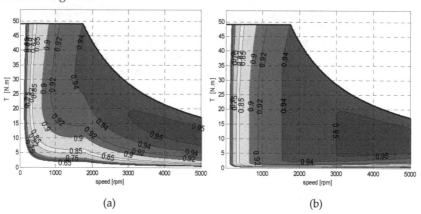

(a)   (b)

Figure 5. Induction motor efficiency maps: a) CF+FW; b) LMA

A complementary perspective is presented in figure 6. It can be seen that the main LMA influence region is below 15 N.m (about 30% of motor nominal torque), with a slight behavior difference, according to n<2000 rpm or n>2000rpm: in the former case (coincident with the constant torque zone), the LMA's efficiency gain is almost constant, while in the late case the energy savings decrease in a smooth way to zero.

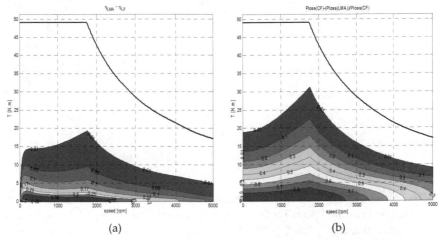

(a)   (b)

Figure 6. LMA Efficiency gain (a) and relative loss differences (b) compared to (CF+FW)

Naturally, LMA acts directly on motor iron losses, since it regulates I_{ds}. However, it has also an impact in motor copper losses, because it provides a better equilibrium between I_{ds} and I_{qs}, particularly in regions where I_{ds} regulation has wide limits. This can be seen in (7).

As a side-note, when considering the plane surfaces in figure 4 (for $I_{d\ max}$ and $I_{d\ min}$), interesting correlations can be made with figure 3, through ($I_{d\ max}$, $I_{d\ min}$) dashed lines and the torque hyperbolas (e.g. higher torques are provided by $I_{d\ min}$ as the speed grows).

4. Simulation model

To evaluate the LMA's contributions to the EV energy consumption reduction, and comparing it to the conventional flux regulation, a simulation study was performed with four diferent driving cycles: ECE-R15, Europe: City, 11-Mode (Japan) and FTP-75. Simulation with other drive-cycles was also implemented, but results achieved with these four give a wide overview of LMA's features. For that purpose, a Matlab/Simulink model was built, which is represented in figure 7.

Basically, I_d^* is generated through (CF+FW) method or by the LMA – blocks (3a) and (3b), respectively. The induction motor is controlled by conventional rotor FOC (block 4); the motor model in block 5 is presented in section 4.4. The motor load and speed references are generated based on a particular drive cycle features (block 1), which includes the vehicle dynamic and mechanical transmission models. Finally, block 2 implements the speed controller (based on a proportional+integral(PI) control law) which generates the motor torque reference. In the following sections, the main model blocks are discribed.

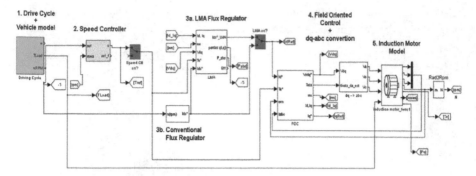

Figure 7. Global simulation model

4.1. Drive cycle+vehicle model

The drive cycles plus the vehicle and mechanical transmission models were implemented with the QuasiStatic Simulation Toolbox (QSS TB), based on Matlab/Simulink, developed by (Guzzella, Amstutz, 2005). The QSS TB library integrates a set of several elements, such as driving cycles, vehicle dynamics, internal combustion engine, electrical motor and mechanical transmisson. Batteries, supercapacitor and fuel cell are also included.

Essentially, it considers a backward (wheel-to-engine) quasi-satic causal model which, based on driving cycle speeds (at discrete times), calculates accelerations and determines the necessary forces, based on the vehicle features and an eventual mechanical transmission. The implemented model includes the QSS TB elements depicted in figure 8.

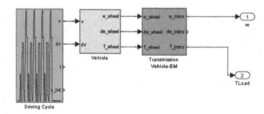

Figure 8. Drive cycle and vehicle/transmission models

The load power demanded to the induction motor (Tload*ωr) considers the drive cycle, vehicle dynamics (rolling and aerodynamic resistance, only in the plane) and also a mechanical transmission with a fixed gear ratio. The vehicle dynamics is modelled by the following equation:

$$M_t \frac{dv(t)}{dt} = F_d(t) - M_t g C_r - \frac{1}{2}\rho C_w A v(t)^2 \tag{33}$$

Where:

M_t - vehicle mass + equivalent mass of rotating parts;

$v(t)$ – vehicle instantaneous longitudinal speed;

$F_d(t)$ – instantaneous driving force;

g – gravity acceleration;

C_r, C_w – rolling friction coefficient, aerodynamic drag coefficient;

ρ, A – air density; vehicle's cross section.

Besides the inertia force, associated to vehicle displacement, the inertia of rotating parts (i.e., kinetic energy stored on it caused by rotational movement) should also be considered, since it is the motor(s) who supply it. This is considered in the "equivalent mass of rotating parts" M_t term (see Table 3). It should be noted that driving cycle block output speed (v) and acceleration (dv) are discrete values. The time step size default value is 1 s; however, in order to increase simulation accuracy, its value was fixed in 0,01 s.

Vehicle and transmission parameters are shown in Tables 3 and 4:

Total vehicle's mass (kg)	350
Rotating mass (%)	5
Vehicle's cross section (m²)	1,5
Wheel diameter (m)	0,3
Aerodynamic drag coefficient	0,3
Rolling friction coefficient	0,008

Table 3. Vehicle Parameters

Gear ratio	5
Efficiency (%)	98
Idling losses by friction (W)	10
Minimum wheel speed beyond which losses are generated (rad/s)	1

Table 4. Mechanical Transmission Parameters

4.2. Rotor flux setpoint generation

a.   LMA

Figure 9 presents the developed LMA block set. R_d and R_q are inputs for the block regions "$w_e < w_n$" and "$w_e > w_n$". Basically, these two elements generate $I_{ds}{}^*$, according to 3.1.3. As it was described, for zones in the (i_{ds}; i_{qs}) plane limited by restrictions (10), (11) and $i_{ds} \leq I_{dn}$, equation (27) is applied. For the border lines, the three defined regions must be considered: in region 1, $I_{ds}{}^*$ is restricted to its maximum allowable value (I_{dn}); for regions 2 and 3, only voltage limit is considered in $I_{ds}{}^*$ generation restriction. Since $T_{p2} = T_{p3}$, the same block can be used for generating $I_{ds}{}^*$ in these two regions.

Since the flux level should not decrease below a minimum value (I_{d_min}), in order to guarantee that $I_{d\_min} \leq I_{ds} \leq I_{dn}$, two saturation blocks are placed at "$w_e < w_n$" and "$w_e > w_n$" outputs.

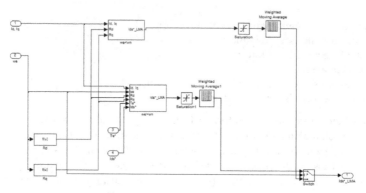

Figure 9. LMA block contents

Figure 10 shows the interior of "we<wn" block.

As it can be seen, $I_{ds}*$ is generated by (27), while $I_{ds}<I_{dn}$; after that $I_{ds}*=I_{dn}$. It should be noted that the absolute value of I_{qs} must be used, in order to consider both motor and braking modes.

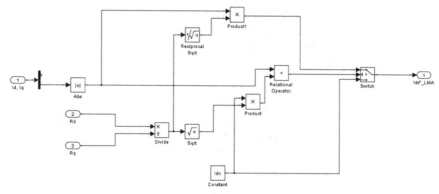

Figure 10. $I_{ds}*$ generation in region 1 (blue path in figure 3)

The block "we>wn" is represented in figure 11. Equation (27) regulates $I_{ds}*$ generation until (31) is no longer true (notice that the absolute value of i_{qs} is compared to the product of "V_{max} restriction" by $(R_d/R_q)^{1/2}$). After that, $I_{ds}*$ is given by zone 3 solution in table 1 (s₃)– "Id* for Vmax restriction border" block. It also should be pointed that when a load point overcomes the voltage limit, the result given by (s₃) is a complex value. In order to deal with this issue, for these situations I_{ds} is taken from the conventional flux regulator.

In contrast with the LMA, the conventional flux regulation (depicted as block (3b) in Figure 7) generates a I_{ds} setpoint according to the following strategy:

$$\begin{cases} I_{ds} = I_{dn} & n \leq n_n \\ I_{ds} = n_n/n \cdot I_{dn} & n > n_n \end{cases} \tag{34}$$

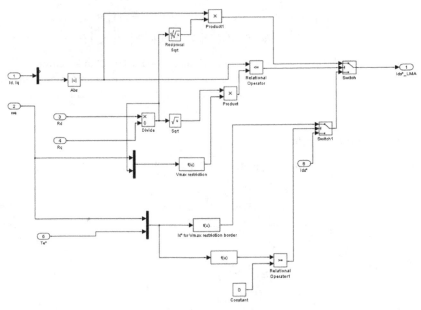

Figure 11. Ids* generation for regions 2 and 3 (green and gray paths in figure 3)

4.3. Rotor indirect FOC

Figure 12 shows the block structure for indirect FOC.

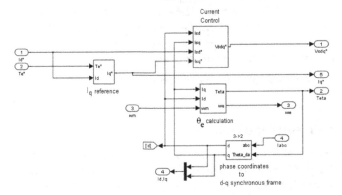

Figure 12. Rotor indirect FOC implementation

Equations (1) and (2) are the basis of "Iq reference" block. Equations (3)-(6) are implemented in "θ_e calculation" block (notice that $\omega_e = \omega_{slip} + \omega_r$). The bottom block considers the coordinates change of instantaneous stator currents, from phase domain to d-q synchronous frame. To do so, the following well known coordinate transformation matrix is applied:

$$\begin{bmatrix} i_{qs} \\ i_{ds} \end{bmatrix} = \frac{2}{3} \begin{bmatrix} \sin\theta_e & \sin(\theta_e - \frac{2}{3}\pi) & \sin(\theta_e + \frac{2}{3}\pi) \\ \cos\theta_e & \cos(\theta_e - \frac{2}{3}\pi) & \cos(\theta_e + \frac{2}{3}\pi) \end{bmatrix} \begin{bmatrix} i_{sa} \\ i_{sb} \\ i_{sc} \end{bmatrix} \tag{35}$$

The "Current Control" block generates stator reference voltage ($V_{sdq}*$) in synchronous frame (through PI's current i_{ds} and i_{qs} controllers), which is applied to the motor model, in phase coordinates, in order to make the real instantaneous stator currents to achieve the reference values.

4.4. Induction motor model

Figure 13 presents the induction motor model considered in simulations, which also includes motor iron losses.

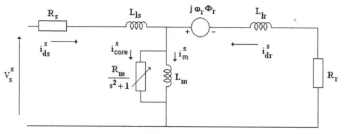

Figure 13. Induction motor model simulated (space vectors in stator reference frame)

When comparing this model to the one considered in LMA (figure 2), the major differences are in parallel (magnetizing) branch. Since core losses currents are not considered in the major circuit, it is expectable that the voltages (V^e_{dm} and V^e_{qm}) on the independent sources are larger compared to the parallel branch voltages in the equivalent model of figure 13. Since core losses are given by $((V^e_{dm})^2+(V^e_{qm})^2)/R_m$, it seems plausible to admit that the core losses in LMA model are higher than the ones in figure 13 model.

Figure 14 shows the simulink implementation of the considered induction motor model (block 5 in figure 7).

5. Simulation results and analysis

An important note is that simulation results were extracted through block 5 (see figure 7), where P_u is obtained directly through $T_e \cdot \omega_r$, based on the drive cycle reference values. In block 3a, Pu is achieved considering $P_{ab}-p_{losses}$ (note that $P_{ab}=u_{sa}i_{sa}+u_{sb}i_{sb}+u_{sc}i_{sc}$, i.e. the sum of instantaneous power of motor phases a, b, c – see figure 14). Motor losses considered by LMA are based on equation (7). There are some differences in P_u values when block 5 or block 3a are considered, which seems to put in evidence the issue mentioned in 4.4.

For each drive cycle, results are presented following the same pattern: the first figure includes the main results for conventional flux regulation and LMA. The second figure

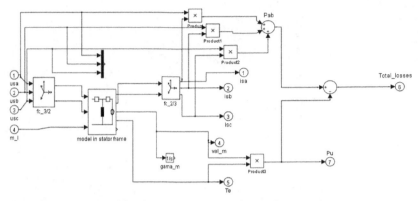

Figure 14. Induction motor model of figure 15 (stator d-q reference frame)

represents the load torque demanded by the drive cycle, while in the third one the motor limits and working points imposed by the drive cycle are illustrated, together with the most significant LMA's efficiency gain zones. Finally, a table with LMA and conventional flux regulation energy performances is also presented.

From a general perspective, these results confirm the main LMA features, described in section 3.2 visible differences from conventional flux regulation occur for low load torque, particularly for relative low speeds. This agrees to the fact that in regions where I_{ds} has a large regulation flexibility, LMA and conventional flux regulation have clearly different performances.

5.1. ECE-R15

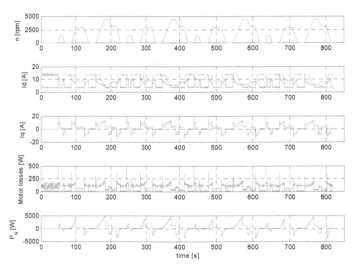

Figure 15. Drive-cycle; (I_d; I_q; Motor losses) – [blue:LMA; red dashed line: conventional regulation]; P_u

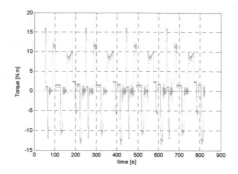

Figure 16. Torques [T_load (green); Te*(red); Te (blue)]

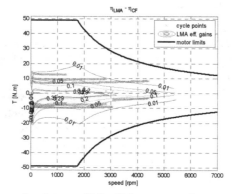

Figure 17. ECE-R15 drive cycle points over LMA efficiency curve gain

	Without LMA	With LMA
Eu (kJ)	221,8	221,5
Eab (kJ)	305,5	266,5
Motor losses (kJ)	83,7	45,0
Energy efficiency (%)	72,6	83,1

Table 5. ECE-R15 energy performances (Eu: energy supplied by the induction motor for the considered drive cycle; Eab: energy absorbed by the induction motor)

In almost 50% of the ECE-R15 drive cycle duration, motor speed is between 0 and 2000 rpm, with the motor torque among -13 Nm and 16 Nm (aprox.) – see figure 16. So, LMA inclusion allows significant loss reductions (table 5): with LMA, total losses are about 20% of E_u (energy supplied by the motor); without LMA, goes up to 38% of E_u. As expected, the main I_{ds} differences occur for n<2000 rpm, particularly for low torques (with LMA, smaller I_{ds} values are clearly visible). In a similar way, LMA performance in braking modes brings good results, since demanded torque has always low values. It should be pointed that when the vehicle is immobilized (I_{qs}=0), LMA performance leads to very significant results (figure 15), since I_{ds} is regulated to its minimum value, while with conventional flux regulation, I_{ds} has its maximum value. In this case, motor iron losses are much higher when compared to the ones with LMA.

From an energy perspective, although LMA acts directly on the iron losses (since it regulates I_{ds}), it has also an impact in motor copper losses (as mentioned in section 3.2). Although for a given torque value, I_{qs} with LMA is higher than with conventional flux regulation (since I_{ds} is smaller), a better equilibrium between I_{ds} and I_{qs} is achieved with LMA. Since copper losses are also dependent on I_{ds}^2 and I_{qs}^2, the motor efficiency has a clear improvement in this drive cycle scenario, which may be seen from figure 17. Nevertheless, efficiency values are relative low, which is no surprise if one take into consideration the efficiency maps (figure 5) together with figure 17 (notice the efficiency (power) curve gains and cycle working points, particularly for n<2000 rpm).

5.2. Europe: City

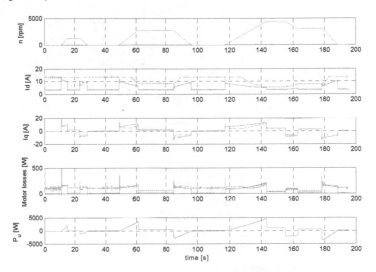

Figure 18. Drive-cycle; (Id; Iq; Motor losses) – [blue:LMA; red dashed line: conventional regulation]; Pu

For about 60% of total time of the "Europe: City" cycle, the vehicle speed is also below 2000 rpm, with the motor torque between -13 Nm and 16 Nm (aprox.). The vehicle is at rest for about 25% of the drive cycle duration. Basically, it puts the motor in the same (T,ω) working region as ECE-R15 (see figures 17 and 20). However, since it has a short time period (195 seg.), energy level demanded is much lower – the lowest one from the chosen drive cycle set. Similar relative energy losses are achieved: 20% for LMA and 36% without LMA of Eu (table 6). In both motor and braking modes, LMA most relevant results are in low speed – low torque region.

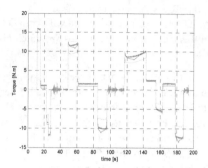

Figure 19. Torques [T_load (green); Te*(red); Te (blue)]

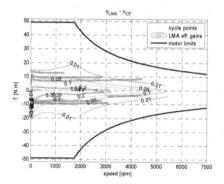

Figure 20. Europe: City drive cycle points over LMA efficiency curve gain

	Without LMA	With LMA
Eu (kJ)	55,5	55,4
Eab (kJ)	75,3	66,6
Motor losses (kJ)	19,8	11,2
Energy efficiency (%)	73,6	83,2

Table 6. Europe: City energy performances

The slightly efficiency increase for this cycle (when compared to ECE-R15) may be associated to the relative decrease of vehicle resting period (about 33% in ECE-R15).

5.3. 11 – Mode (Japan)

Drive cycle period where n<2000 rpm is relative short (<33%); the motor torque lies between 13 Nm and -10 Nm and the vehicle is immobilized a little less than 25% of the cycle duration. As expected, it's in the initial resting time period and on the final 25 sec that I_{ds} values generated by the LMA are significantly different from the conventional flux I_{ds}

values. In other words, the motor losses difference are attached to these time periods, particularly to the resting one (figure 21). On the other drive cycle periods, motor losses are very similar (notice that in some intervals, LMA losses are slightly larger. This unexpected result is most probably related to the issue discussed in section 4.4).

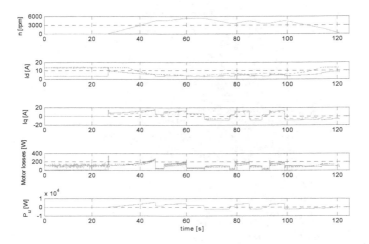

Figure 21. Drive-cycle; (Id; Iq; Motor losses) – [blue:LMA; red dashed line: conventional regulation]; P_u

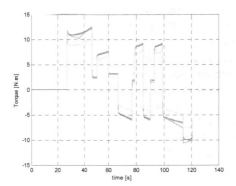

Figure 22. Torques [T_load (green); Te*(red); Te (blue)]

LMA total losses are about 14% of E_u, while conventional regulation losses are 16% of E_u. Although curve efficiency gains in figure 23 are referred to power efficiency, cycle working points somehow agree with efficiency energy gain achieved with LMA (table 7): for n> 2000 rpm there is a significant number of points between 1% and 5 % efficiency curves gain; also notice that some points are below 1% efficiency gain.

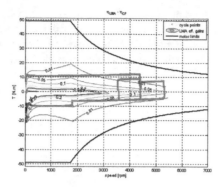

Figure 23. 11-Mode drive cycle points over LMA efficiency curve gain

	Without LMA	With LMA
Eu (kJ)	76,7	76,7
Eab (kJ)	89,3	87,5
Motor losses (kJ)	12,5	10,8
Energy efficiency (%)	86	87,6

Table 7. 11-Mode energy performances

5.4. FTP-75

For this drive cycle, the time period for which n<2000 rpm is shorter then the previous cycles. Motor torque limit is now -20 Nm and 25 Nm (aprox), while maximum speed is 8000 rpm. Frequent accelerations, as well as its long time period (1840 sec), make this cycle the most energy demanding. At the same time, pushes the motor to its limits: figure 24 shows that motor exceeds its nominal power between [200-300] s and later in the interval [1500-1700] sec. However, this overload (whose maximum instantaneous power is about 11 kW) occurs for a small number of intervals, each one with a very short existence. This way, it's reasonable to assume that motor is not under electric hazardous working conditions. From a mechanical perspective, maximum speed - about 4 times motor nominal speed – is reached for relative short intervals, so one may assume that the motor (and the vehicle) will be safe in this working conditions.

Due to high speeds and relative high torque demand (figure 25), LMA shows a relative performance closer to the conventional regulation. As expected, relevant differences for I_d generation occur for relative low speed (basically, when the vehicle is at rest) and low torque values, e.g. intervals [50-200; 800-950] sec. – figures 25 and 26. With LMA and without it, motor losses are, respectively, 11,4% and 13,1% of E_u (table 8). Motor efficiency map (figure 5) explains the high efficiency values associated to this cycle, while the small efficiency gain achieved is according to figures 25 and 26.

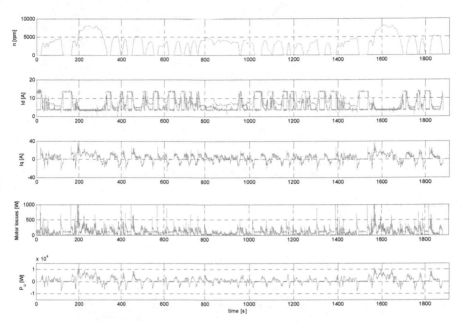

Figure 24. Drive-cycle; (Id*-red & Id-blue); (Iq*-red & Iq-blue); P_{ab} and motor losses (without LMA)

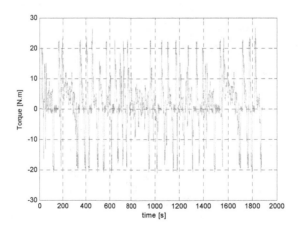

Figure 25. Torques [T_load (green); Te*(red); Te (blue)]

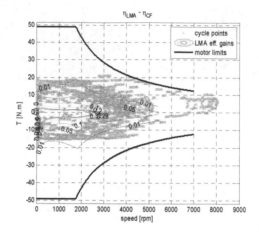

Figure 26. FTP-75 drive cycle points over LMA efficiency curve gain

	Without LMA	With LMA
Eu (kJ)	1716	1716
Eab (kJ)	1941	1910
Motor losses (kJ)	225,1	194,6
Energy efficiency (%)	88,4	89,8

Table 8. FTP-75 energy performances

Figures 27 and 28 present, respectively, induction motor energy consumption, efficiency and losses for each simulated drive cycle, with and without LMA.

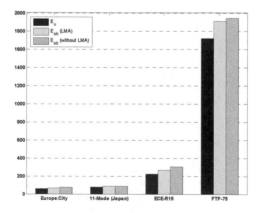

Figure 27. Drive-Cycles energy consumptions [kJ]

As a final remark, it is interesting to note that Europe:city and ECE-R15 have similar efficiency levels; also the same fact can be seen for 11-Mode and FTP-75.

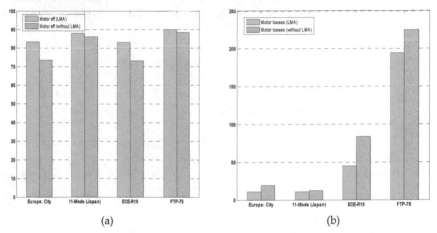

(a)   (b)

Figure 28. Drive-Cycles energy efficiency [%] (a) and motor losses [kJ] (b)

6. Conclusion

Induction motor drives for EVs are submitted to a large set of working conditions, quite different from rated ones. Motor energy saving is fundamental for improving EVs performances. Under the loss model based approach previously discussed (LMA), a set of simulation results was presented in this book chapter, aiming to improve the induction motor energy performance. Different standard driving cycle scenarios were considered in order to evaluate the chosen LMA features: compared to conventional flux regulation, the major improvements in motor efficiency are for low load torque, particularly for relative low speeds. These are the motor working points where its efficiency is tipically lower, which is an interesting LMA feature. This is in agreement to the fact that LMA action has a more significative impact on ECE-R15 and Europe:city efficiencies, as explained through figures 15,17 and 18, 20 analysis.

Due to LMA impact on iron losses (function of I_d), a possibility to be considered in future works is the impact of LMA on motors with higher power rates and/or high efficiency level motors, where the relative weights of iron and copper losses are different.

Author details

Pedro Melo
Polytechnic Institute of Porto, Portugal

Ricardo de Castro and Rui Esteves Araújo
Faculty of Engineering – University of Porto, Portugal

7. References

Bazzi, A., & Krein, P. (2010). Review of Methods for Real-Time Loss Minimization in Induction Machines. *IEEE Transactions on Industrial Applications*, Vol.41, No.6, pp. 2319-2328.

Fernandez-Bernal, F., Garcia-Cerrada, A., & Faure, R. (2000). Model-based loss minimization for DC and AC vector-controlled motors including core saturation. *IEEE Transactions on Industrial Applications*, Vol.36, No.3, pp. 755-763.

Garcia, G., Luis, J., Stephan, R., & Watanabe, E. (1994). An efficient controller for an adjustable speed induction motor drive. *IEEE Transactions on Industrial Electronics*, Vol.41, No.5, pp. 533-539.

Guzzella, L., & Amstutz, A. (2005), *The QSS Toolbox Manual*, Measurement and Control Laboratory –Swiss Federal Institute of Technology Zurich.

Kioskeridis, I., & Margaris, N. (1996). Loss minimization in induction motor adjustable-speed drives. *IEEE Transactions on Industrial Electronics*, Vol.43, No.1, pp. 226-231.

Krishnan, R. (2001). *Electric Motor Drives – Modeling, Analysis and Control* (1 edition), Prentice Hall, ISBN 13: 978-0130910141.

Lim, S. and K. Nam (2004). Loss-minimising control scheme for induction motors. *IEE Proceedings - Electric Power Applications*, Vol.151, No.4, pp. 385-397.

Novotny, D., Lipo, T. (1996). *Vector control and dynamics of AC drives*, Clarendon Press, Oxford, ISBN 9780198564393.

Zeraoulia, M., M. E. H. Benbouzid, et al. (2006). Electric Motor Drive Selection Issues for HEV Propulsion Systems: A Comparative Study. *IEEE Transactions on Vehicular Technology*, Vol.55, No.6, pp. 1756-1764.

Chapter 6

The Asymmetrical Dual Three-Phase Induction Machine and the MBPC in the Speed Control

Raúl Igmar Gregor Recalde

Additional information is available at the end of the chapter

1. Introduction

Recent research has focused on exploring the advantages of multiphase[1] machines over conventional three-phase systems, including lower torque pulsations, less DC-link current harmonics, higher overall system reliability, and better power distribution per phase [1]. Among these multiphase drives, the asymmetrical dual three-phase machines with two sets of three-phase stator windings spatially shifted by 30 electrical degrees and isolated neutral points is one of the most widely discussed topologies and found industrial application in more-electric aircraft, electrical and hybrid vehicles, ship propulsion, and wind power systems [2]. This asymmetrical dual three-phase machines is a continuous system which can be described by a set of differential equations. A methodology that simplifies the modeling is based on the vector space decomposition (VSD) theory introduced in [3] to transform the original six-dimensional space of the machine into three two-dimensional orthogonal subspaces in stationary reference frame $(\alpha - \beta)$, $(x - y)$ and $(z_1 - z_2)$. From the VSD approach, can be emphasized that the electromechanical energy conversion variables are mapped in the $(\alpha - \beta)$ subspace, meanwhile the current components in the $(x - y)$ subspace represent supply harmonics of the order $6n \pm 1$ $(n = 1, 3, 5, ...)$ and only produce losses, so consequently should be controlled to be as small as possible. The voltage vectors in the $(z_1 - z_2)$ are zero due to the separated neutral configuration of the machine, therefore this subspace has no influence on the control [4].

Model-based predictive control (MBPC) and multiphase drives have been explored together in [5, 6], showing that predictive control can provide enhanced performance for multiphase drives. In [7, 8], different variations of the predictive current control techniques are proposed to minimize the error between predicted and reference state variables, at the expense of increased switching frequency of the insulated-gate bipolar transistor (IGBTs). On the other hand are proposed control strategies based on sub-optimal solutions restricted the available voltage vectors for multiphase drive applications aiming at reducing the computing cost and

---

[1] The multiphase term, regards more than three phase windings placed in the same stator of the electric machine.

improving the drive performance [9]. This chapter wide the concept of the MBPC techniques to the speed control of a dual three-phase induction machine, by using an Kalman Filter (KF) to improve the estimation of states through an optimal estimation of the rotor current. The KF is an efficient recursive filter that estimates the internal state of a dynamic system from a series of noisy measurements. Its purpose is to use measurements that are observed over time that contain noise (random variations) and other inaccuracies (including modeling errors), and produce values that tend to be closer to the true values of the measurements and their associated calculated values. This feature is an attractive solution in the predictive control of induction machines based on the model, mainly if not precisely known internal parameters of the drive, and the measurement of the state variables are perturbed by gaussian noise.

The chapter includes simulation results of the current control based on a predictive model of the asymmetrical dual three-phase induction machine and proposes a new approach to speed control based on MBPC technique. The results provided confirm the feasibility of the speed control scheme for multi-phase machines. The rest of the chapter is organized as follows. Section 2 introduces an asymmetrical dual three-phase AC drive used for simulations. Section 3 details the general principles of the predictive current control method for AC drives. Section 4 shows the simulation results obtained from the inner loop of predictive current control and proposed a new approach to speed control for the dual three-phase induction machine, on the other hand presents a discussion of the obtained results from the proposed approach. The chapter ends with Section 5 where the conclusions are presented.

2. The asymmetrical dual three-phase AC drive

The asymmetrical dual three-phase induction machine is supplied by a 6-phase voltage source inverter (VSI) and a Dc Link, as shown in Figure 1. This six-phase machine is a continuous system which can be described by a set of differential equations. A methodology that simplifies the modeling is based on the vector space decomposition (VSD) theory introduced in [3] to transform the original six-dimensional space of the machine into three

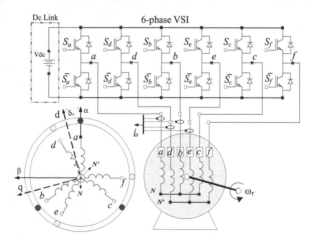

Figure 1. A general scheme of an asymmetrical dual three-phase drive

two-dimensional orthogonal subspaces in stationary reference frame $(\alpha - \beta)$, $(x - y)$ and $(z_1 - z_2)$, by means of a 6×6 transformation matrix using an amplitude invariant criterion:

$$
\mathbf{T} = \frac{1}{3}
\begin{bmatrix}
1 & \frac{\sqrt{3}}{2} & -\frac{1}{2} & -\frac{\sqrt{3}}{2} & -\frac{1}{2} & 0 \\
0 & \frac{1}{2} & \frac{\sqrt{3}}{2} & \frac{1}{2} & -\frac{\sqrt{3}}{2} & -1 \\
1 & -\frac{\sqrt{3}}{2} & -\frac{1}{2} & \frac{\sqrt{3}}{2} & -\frac{1}{2} & 0 \\
0 & \frac{1}{2} & -\frac{\sqrt{3}}{2} & \frac{1}{2} & \frac{\sqrt{3}}{2} & -1 \\
1 & 0 & 1 & 0 & 1 & 0 \\
0 & 1 & 0 & 1 & 0 & 1
\end{bmatrix}
\tag{1}
$$

The VSI has a discrete nature and has a total number of $2^6 = 64$ different switching states defined by six switching functions corresponding to the six inverter legs $[S_a, S_d, S_b, S_e, S_c, S_f]$, where $S_i \in \{0,1\}$. The different switching states and the voltage of the DC link (Vdc) define the phase voltages which can in turn be mapped to the $(\alpha - \beta) - (x - y)$ space according to the VSD approach. Consequently, the 64 different on/off combinations of the six VSI legs lead to 64 space vectors in the $(\alpha - \beta)$ and $(x - y)$ subspaces. Figure 2 shows the active vectors in the $(\alpha - \beta)$ and $(x - y)$ subspaces, where each vector switching state is identified using the switching function by two octal numbers corresponding to the binary numbers $[S_a S_b S_c]$ and $[S_d S_e S_f]$, respectively. For the sake of conciseness, the 64 VSI switching vectors will be usually referred as voltage vectors, or just vectors, in what follows. It must be noted that the 64 possibilities imply only 49 different vectors in the $(\alpha - \beta) - (x - y)$ space. Nevertheless, redundant vectors should be considered as different vectors because they have a different impact on the switching frequency even though they generate identical torque and losses in the six-phase machine.

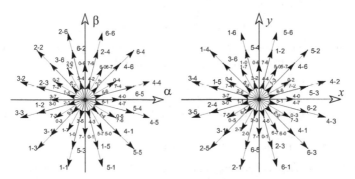

Figure 2. Voltage vectors and switching states in the $(\alpha - \beta)$ and $(x - y)$ subspaces for a 6-phase asymmetrical VSI

To represent the stationary reference frame $(\alpha - \beta)$ in dynamic reference $(d - q)$, a rotation transformation must be used. This transformation is given by:

$$
\mathbf{T}_{dq} =
\begin{bmatrix}
\cos(\delta_r) & -\sin(\delta_r) \\
\sin(\delta_r) & \cos(\delta_r)
\end{bmatrix}
\tag{2}
$$

where δ_r is the rotor angular position referred to the stator as shown in Figure 1.

2.1. Machine model in the $(\alpha - \beta)$ subspace

The asymmetrical dual three-phase machine model can be obtained using a specific and convenient choice of state-space variables, for example, stator and rotor currents. Thus the six-phase machine can be modelled in a stationary reference frame according to the VSD approach as:

$$[u]_{\alpha\beta} = [G] \frac{d}{dt} [x]_{\alpha\beta} + [F] [x]_{\alpha\beta} \tag{3}$$

$$[u]_{\alpha\beta} = \left[u_{\alpha s} \ u_{\beta s} \ 0 \ 0 \right]^T ; [x]_{\alpha\beta} = \left[i_{\alpha s} \ i_{\beta s} \ i_{\alpha r} \ i_{\beta r} \right]^T \tag{4}$$

where $[u]_{\alpha\beta}$ is the input vector, $[x]_{\alpha\beta}$ is the state vector and $[F]$ and $[G]$ are matrices that define the dynamics of the electrical drive that for this set of state variables are:

$$[F] = \begin{bmatrix} R_s & 0 & 0 & 0 \\ 0 & R_s & 0 & 0 \\ 0 & \omega_r \cdot L_m & R_r & \omega_r \cdot L_r \\ -\omega_r \cdot L_m & 0 & -\omega_r \cdot L_r & R_r \end{bmatrix} ; [G] = \begin{bmatrix} L_s & 0 & L_m & 0 \\ 0 & L_s & 0 & L_m \\ L_m & 0 & L_r & 0 \\ 0 & L_m & 0 & L_r \end{bmatrix} \tag{5}$$

where ω_r is the rotor angular speed, and the electrical parameters of the machine are the stator and rotor resistances R_s, R_r, the stator and rotor inductances $L_s = L_{ls} + L_m$, $L_r = L_{lr} + L_m$, the stator and rotor leakage inductances L_{ls}, L_{lr} and the magnetization inductance L_m. Using selected state-space variables and amplitude invariant criterion in the transformation, the mechanical part of the drive is given by the following equations:

$$T_e = 3.P \left(\psi_{\beta r} i_{\alpha r} - \psi_{\alpha r} i_{\beta r} \right) \tag{6}$$

$$J_i \frac{d}{dt} \omega_r + B_i \omega_r = P \left(T_e - T_L \right) \tag{7}$$

where T_e is the generated torque, T_L the load torque, P the number of pair of poles, J_i the inertia coefficient, B_i the friction coefficient and $\psi_{\alpha\beta r}$ the rotor flux.

The $(\alpha - \beta)^2$ axes are selected in such a manner that they coincide with the plane of rotation of the airgap flux. Therefore, these variables will are associated with the production of the airgap flux in the machine and with the electromechanical energy conversion related [3].

2.2. Machine model in the $(x - y)$ subspace

Because the $(x - y)$ subspace is orthogonal to the $(\alpha - \beta)$ subspace, the projected variables in this subspace will do not contribute to the airgap flux, and therefore are not related to energy conversion. This model are limited only by the stator resistance and stator leakage inductance, as shown in the following equation:

$$\begin{bmatrix} u_{xs} \\ u_{ys} \end{bmatrix} = \begin{bmatrix} L_{ls} & 0 \\ 0 & L_{ls} \end{bmatrix} \frac{d}{dt} \begin{bmatrix} i_{xs} \\ i_{ys} \end{bmatrix} + \begin{bmatrix} R_s & 0 \\ 0 & R_s \end{bmatrix} \begin{bmatrix} i_{xs} \\ i_{ys} \end{bmatrix} \tag{8}$$

---

[2] It can be noted that $(\alpha - \beta)$ equations are similar to those of a three-phase machine while that, as will be seen in the following section, the $(x - y)$ equations do not link the rotor side and consequently do not influence the machine dynamics but are source of Joule losses in the machine.

3. Predictive model

The machine model must be discretized in order to be of use as a predictive model. Taking into account that the electromechanical energy conversion involves only quantities in the $(\alpha - \beta)$ subspace, the predictive model could be simplified, discarding the $(x - y)$ subspace. Assuming the asymmetrical dual three-phase induction machine model (see Equation 3) and using the following state components $(x_1 = i_{\alpha s}, \; x_2 = i_{\beta s}, \; x_3 = i_{\alpha r}, \; x_4 = i_{\beta r})$, the resulting equations can be written as:

$$\dot{x}_1 = c_3 \left(R_r x_3 + \omega_r x_4 L_r + \omega_r x_2 L_m \right) + c_2 \left(u_{\alpha s} - R_s x_1 \right)$$
$$\dot{x}_2 = c_3 \left(R_r x_4 - \omega_r x_3 L_r - \omega_r x_1 L_m \right) + c_2 \left(u_{\beta s} - R_s x_2 \right)$$
$$\dot{x}_3 = c_4 \left(-R_r x_3 - \omega_r x_4 L_r - \omega_r x_2 L_m \right) + c_3 \left(-u_{\alpha s} + R_s x_1 \right)$$
$$\dot{x}_4 = c_4 \left(-R_r x_4 + \omega_r x_3 L_r + \omega_r x_1 L_m \right) + c_3 \left(-u_{\beta s} + R_s x_2 \right) \tag{9}$$

where c_1-c_4 are constant coeficients defined as:

$$c_1 = L_s \cdot L_r - L_m^2, \; c_2 = \frac{L_r}{c_1}, \; c_3 = \frac{L_m}{c_1}, \; c_4 = \frac{L_s}{c_1} \tag{10}$$

Stator voltages are related to the control input signals through the inverter model. The simplest model has been selected for this case study for the sake of speeding up the optimization process. Then if the gating signals are arranged in vector $\mathbf{S} = \left[S_a, S_d, S_b, S_e, S_c, S_f \right] \in \mathbf{R}^6$, with $\mathbf{R} = \{0,1\}$ the stator voltages are obtained from:

$$\mathbf{M} = \frac{1}{3} \begin{bmatrix} 2 & 0 & -1 & 0 & -1 & 0 \\ 0 & 2 & 0 & -1 & 0 & -1 \\ -1 & 0 & 2 & 0 & -1 & 0 \\ 0 & -1 & 0 & 2 & 0 & -1 \\ -1 & 0 & -1 & 0 & 2 & 0 \\ 0 & -1 & 0 & -1 & 0 & 2 \end{bmatrix} \cdot \mathbf{S}^T \tag{11}$$

An ideal inverter converts gating signals to stator voltages that can be projected to $(\alpha - \beta)$ and $(x - y)$ axes and gathered in a row vector $\mathbf{U}_{\alpha\beta xys}$ computed as:

$$\mathbf{U}_{\alpha\beta xys} = \left[u_{\alpha s}, u_{\beta s}, u_{xs}, u_{ys}, 0, 0 \right]^T = Vdc \cdot \mathbf{T} \cdot \mathbf{M} \tag{12}$$

being Vdc the Dc Link voltage and superscript $(^T)$ indicates the transposed matrix. Combining Equations 9-12 a nonlinear set of equations arises that can be written in state space form:

$$\mathbf{X}(t) = f\left(\mathbf{X}(t), \mathbf{U}(t) \right)$$
$$\mathbf{Y}(t) = \mathbf{C} \mathbf{X}(t) \tag{13}$$

with state vector $\mathbf{X}(t) = [x_1, x_2, x_3, x_4]^T$, input vector $\mathbf{U}(t) = \left[u_{\alpha s}, u_{\beta s} \right]$, and output vector $\mathbf{Y}(t) = [x_1, x_2]^T$. The components of vectorial function f and matrix $\mathbf{C}$ are obtained in a straightforward manner from Equation 9 and the definitions of state and output vector.

Model (Equation 13) must be discretized in order to be of use for the predictive controller. A forward Euler method is used to keep a low computational burden. As a consequence the resulting equations will have the needed digital control form, with predicted variables depending just on past values and not on present values of variables. This leads to the following equations:

$$\hat{\mathbf{X}}(k+1|k) = \mathbf{X}(k) + T_m f\left(\mathbf{X}(k), \mathbf{U}(k)\right)$$
$$\mathbf{Y}(k) = \mathbf{C}\mathbf{X}(k) \tag{14}$$

denoting by (k) the current sample, T_m the sampling time and being $\hat{\mathbf{X}}(k+1|k)$ a prediction of the future next-sample state made at sample time (k).

3.1. Kalman Filter design

Kalman Filter is an optimal recursive estimation algorithm based on the state-space concepts and suitable for digital computer implementation. That is, it is an optimal estimator for computing the conditional mean and covariance of the probability distribution of the state of a linear stochastic system with uncorrelated gaussian process and measurement noise. The algorithm minimizes the estimate error of the states by utilizing knowledge of system and measurements dynamic, assuming statistics of system noises and measurement errors, considering initial condition information [10]. Considering uncorrelated gaussian process and measurement noise, Equations 14 can be written as:

$$\hat{\mathbf{X}}(k+1|k) = \mathbf{A}\mathbf{X}(k) + \mathbf{B}\mathbf{U}(k) + \mathbf{H}\omega(k)$$
$$\mathbf{Y}(k) = \mathbf{C}\mathbf{X}(k) + v(k) \tag{15}$$

the matrices **A**, **B** and **C** are obtained in a straightforward manner from Equation 14 and the definitions of state and output vector, **H** is the noise-weight matrix, $\omega(k)$ is the process noise matrix, and $v(k)$ is the measurement noise matrix. The covariance matrices R_ω and R_v of these noises are defined in function to the expected value $E\{\cdot\}$ as:

$$R_\omega = cov(\omega) = E\left\{\omega \cdot \omega^T\right\}; R_v = cov(v) = E\left\{v \cdot v^T\right\} \tag{16}$$

3.1.1. Reduced-order state estimation

In the state space description of Equation 14 only stator currents, voltages and mechanical speed are measured. Stator voltages are easily predicted from gating commands issued to the VSI, rotor current, however, is not directly measured. This difficulty can be overcome by means of estimating the rotor current using the concept of reduced-order estimators. The reduced-order estimator provide an estimate for only the unmeasured part of state vector, then, the evolution of states can be written as:

$$\begin{bmatrix} \mathbf{X}_a(k+1|k) \\ \cdots \\ \mathbf{X}_b(k+1|k) \end{bmatrix} = \begin{bmatrix} \overline{\mathbf{A}}_{11} & \vdots & \overline{\mathbf{A}}_{12} \\ \cdots & \cdots & \cdots \\ \overline{\mathbf{A}}_{21} & \vdots & \overline{\mathbf{A}}_{22} \end{bmatrix} \begin{bmatrix} \mathbf{X}_a(k) \\ \cdots \\ \mathbf{X}_b(k) \end{bmatrix} + \begin{bmatrix} \overline{\mathbf{B}}_1 \\ \cdots \\ \overline{\mathbf{B}}_2 \end{bmatrix} \mathbf{U}_{\alpha\beta s}(k)$$

$$\mathbf{Y}(k) = \begin{bmatrix} \overline{\mathbf{I}} & \vdots & \overline{\mathbf{0}} \end{bmatrix} \begin{bmatrix} \mathbf{X}_a(k) \\ \cdots \\ \mathbf{X}_b(k) \end{bmatrix} \tag{17}$$

where $\mathbf{I}$ is the identity matrix of order 2x2, $\mathbf{X}_a = \left[i_{\alpha s}(k) \; i_{\beta s}(k) \right]^T$ is the portion directly measured, which is $\mathbf{Y}(k)$, $\mathbf{X}_b = \left[i_{\alpha r}(k) \; i_{\beta r}(k) \right]^T$ is the remaining portion to be estimated, and $\overline{\mathbf{A}}$ and $\overline{\mathbf{B}}$ are matrices obtained in a straightforward manner from Equation 15 and are represented according to the following matrices:

$$
\overline{\mathbf{A}} = \begin{bmatrix}
(1 - T_m.c_2.R_s) & T_m.c_3.L_m.\omega_r & \vdots & T_m.c_3.R_r & T_m.c_3.L_r.\omega_r \\
-T_m.c_3.L_m.\omega_r & (1 - T_m.c_2.R_s) & \vdots & -T_m.c_3.L_r.\omega_r & T_m.c_3.R_r \\
\cdots & \cdots & \cdots & \cdots & \cdots \\
T_m.c_3.R_s & -T_m.c_4.L_m.\omega_r & \vdots & (1 - T_m.c_4.R_r) & -T_m.c_4.L_r.\omega_r \\
T_m.c_4.L_m.\omega_r & T_m.c_3.R_s & \vdots & T_m.c_4.L_r.\omega_r & (1 - T_m.c_4.R_r)
\end{bmatrix}
$$

$$
\overline{\mathbf{B}} = \begin{bmatrix}
T_m.c_2 & 0 \\
0 & T_m.c_2 \\
\cdots & \cdots \\
-T_m.c_3 & 0 \\
0 & -T_m.c_3
\end{bmatrix} \tag{18}
$$

The portion describing the dynamics of the unmeasured states can be written as:

$$
\mathbf{X}_b(k+1|k) = \overline{\mathbf{A}}_{22}\mathbf{X}_b(k) + \overline{\mathbf{A}}_{21}\mathbf{X}_a(k) + \overline{\mathbf{B}}_2\mathbf{U}_{\alpha\beta s}(k) \tag{19}
$$

where the last two terms on the right are known and can be considered as an input into the $\mathbf{X}_b$ dinamics. The $\mathbf{X}_a$ portion may be expressed as:

$$
\mathbf{X}_a(k+1|k) - \overline{\mathbf{A}}_{11}\mathbf{X}_a(k) - \overline{\mathbf{B}}_1\mathbf{U}_{\alpha\beta s}(k) = \overline{\mathbf{A}}_{12}\mathbf{X}_b(k) \tag{20}
$$

Note in Equation 20 that this equation represent a relationship between a measured quantity on the left and the unknown state vector on the right. Therefore, the dynamics of the reduced-order estimator ecuations are:

$$
\hat{\mathbf{X}}_b(k+1|k) = (\overline{\mathbf{A}}_{22} - \mathbf{K}_e\overline{\mathbf{A}}_{12})\hat{\mathbf{X}}_b(k) + \mathbf{K}_e\mathbf{Y}(k+1) +
$$
$$
(\overline{\mathbf{A}}_{21} - \mathbf{K}_e\overline{\mathbf{A}}_{11})\mathbf{Y}(k) + (\overline{\mathbf{B}}_2 - \mathbf{K}_e\overline{\mathbf{B}}_1)\mathbf{U}_{\alpha\beta s}(k) \tag{21}
$$

where, $\mathbf{K}_e$ represents the KF gain matrix based on the covariance of the noise.

3.1.2. Kalman Filter gain matrix evaluation

The KF gain matrix ($\mathbf{K}_e$) is recalculated at each sampling time recursive manner as:

$$
\mathbf{K}_e(k) = \mathbf{\Gamma}(k) \cdot \mathbf{C}^T R_\nu \tag{22}
$$

where $\mathbf{\Gamma}$ is the covariance of the new estimate and is a function of the old estimate covariance (φ) as follows:

$$
\mathbf{\Gamma}(k) = \varphi(k) - \varphi(k) \cdot \mathbf{C}^T (\mathbf{C} \cdot \varphi(k) \cdot \mathbf{C}^T + R_\nu)^{-1} \cdot \mathbf{C} \cdot \varphi(k) \tag{23}
$$

From the state equation which includes the process noise it is possible to obtain a correction of the covariance of the state estimate as:

$$
\varphi(k+1) = \mathbf{A} \cdot \mathbf{\Gamma}(k) \cdot \mathbf{A}^T + \mathbf{H} \cdot R_\omega \cdot \mathbf{H}^T \tag{24}
$$

This completes the required relations for the optimal state estimation. Thus $\mathbf{K}_e$ provides the minimum estimation errors, given a knowledge of the process noise magnitude (R_ω), the measurement noise magnitude (R_v), and the covariance initial condition ($\varphi(0)$) [11].

3.2. Current control loop

The current control loop, based on the MBPC technique avoids the use of modulation techniques since a single switching vector is applied during the whole switching period. The MBPC technique selects the control actions through solving an optimization problem at each sampling period. A model of the real system, is used to predict its output. This prediction is carried out for each possible output, or switching vector, of the six-phase inverter to determine which one minimizes a defined cost function. The proposed scheme is shown in Figure 3.

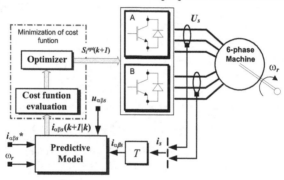

Figure 3. Current control loop based on the MBPC technique

3.2.1. Cost function

The cost function should include all aspects to be optimized. In the current predictive control applied to the asymmetrical dual three-phase induction machine, the most important feature to be optimized are the tracking errors of the stator currents in the $(\alpha - \beta)$ subspace for a next sampling time, since this variables are related to the electromechanical conversion. To minimize the prediction errors at each sampling time k it enough utilize a simple term as:

$$J = \| \hat{e}_{i\alpha s}(k+1|k) + \hat{e}_{i\beta s}(k+1|k) \|^2 \leftrightarrow \begin{cases} \hat{e}_{i\alpha s}(k+1|k) = i_{\alpha s}^*(k+1) - \hat{i}_{\alpha s}(k+1|k) \\ \hat{e}_{i\beta s}(k+1|k) = i_{\beta s}^*(k+1) - \hat{i}_{\beta s}(k+1|k) \end{cases} \quad (25)$$

where $\| \, . \, \|$ denotes the vector modulus, i_s^* is a vector containing the reference for the stator currents and $\hat{i}_s(k+1|k)$ is the prediction of the stator currents calculated from measured and estimated states and the voltage vector $U_{\alpha\beta s}(k)$ as shown in Equation 20. Figure 4 (a) shows the all projection of the stator current predictions calculated from the prediction model. The current control selects the control vector that minimizes the cost function at each sampling time. Figure 4 (b) shows the selection of the optimal vector based on a minimization of prediction errors.

More complicated cost functions can be devised for instance to minimize harmonic content, VSI switching losses, torque and flux and/or active and reactive power. Also, in multi-phase

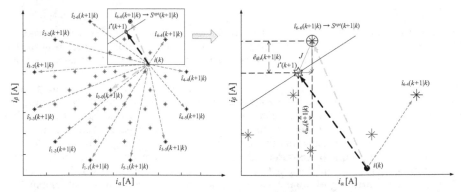

(a) Projection of the stator currents prediction in stationary reference frame $(\alpha - \beta)$

(b) Evaluation of the cost function (J) and selection of the optimal vector (S^{opt})

Figure 4. Minimization of tracking error in stator currents in stationary reference frame $(\alpha - \beta)$

drives stator current can be decomposed in subspaces in different ways. An appropriate decomposition allow to put more emphasis on harmonic reduction as will be shown in the case study for a six-phase motor drive [5, 12]. The more relevant cost functions are shown in Table 1. The superscript $(^*)$ denote the reference value and the terms involved in each cost function are detailed in the Table 2.

Controlled variables	Cost functions (J)
Currents $(\alpha\text{-}\beta)$ and harmonic $(x - y)$	$||i_\alpha^* - i_\alpha| + |i_\beta^* - i_\beta||^2 + \lambda.||i_x^* - i_x| + |i_y^* - i_y||$
Active and reactive power	$|Q_{in}| + |P_{in}^* - P_{in}|$
Torque and flux	$|T_e^* - T_e| + \lambda||\psi_s^*| - |\psi_s||$
Currents $(\alpha\text{-}\beta)$ and voltage balance	$||i_\alpha^* - i_\alpha| + |i_\beta^* - i_\beta| + \lambda.|V_{c1} - V_{c2}|$
Currents $(\alpha\text{-}\beta)$ and VSI switching losses	$||i_\alpha^* - i_\alpha| + |i_\beta^* - i_\beta|| + \lambda.N_s$

Table 1. Possible cost functions in function to the controlled variables

Variable	Description
i_α	Measured α current
i_β	Measured β current
i_x	Measured x current
i_y	Measured y current
Q_{in}	Reactive power
P_{in}	Active power
T_e	Torque
ψ_s	Flux of the stator
λ	Weighting factor
V_{c1}, V_{c2}	Voltages on each capacitor (VSI balanced)
N_s	Number of switches

Table 2. Description of the terms involved in each cost function of the Table 1

3.2.2. Optimizer

The predictive model should be used 64 times to consider all possible voltage vectors. However, the redundancy of the switching states results in only 49 different vectors (48 active and 1 null) as shown on Figure 2. This consideration is commonly known as the optimal solution. The number of voltage vectors to evaluate the predictive model can be further reduced if only the 12 outer vectors (the largest ones) are considered. This assumption is commonly used if sinusoidal output voltage is required and it is not necessary to synthesize $(x - y)$ components. In this way, the optimizer can be implemented using only 13 possible stator voltage vectors[3]. This way of proceeding increases the speed at which the optimizer can be run, allowing decreasing the sampling time at the cost of losing optimality. A detailed study of the implications of considering the optimal solution can be found at [6]. For a generic multi-phase machine, where f is the number of phase and ε the search space (49 or 13 vectors), the control algorithm proposed produces the optimum gating signal combination S^{opt} as follows:

Algorithm 1 Proposed algorithm

---

  comment: Compute the covariance matrix. Equation 23
  $\Gamma(k) = \varphi(k) - \varphi(k) \cdot \mathbf{C}^T (\mathbf{C} \cdot \varphi(k) \cdot \mathbf{C}^T + R_v)^{-1} \cdot \mathbf{C} \cdot \varphi(k)$
  comment: Compute the KF gain matrix. Equation 22
  $\mathbf{K}_e(k) = \Gamma(k) \cdot \mathbf{C}^T R_v$
  comment: Optimization algorithm
  $J_o := \infty, i := 1$
  while $i \leq \varepsilon$ **do**
    $\mathbf{S}_i \leftarrow \mathbf{S}_{i,j} \, \forall j = 1, ..., f$
    comment: Compute stator voltages. Equation 12
    $U_{\alpha\beta xys} = \left[u_{\alpha s}, u_{\beta s}, u_{xs}, u_{ys}, 0, 0 \right]^T = Vdc \cdot \mathbf{T} \cdot \mathbf{M}$
    comment: Compute a prediction of the state. Equation 15
    $\hat{\mathbf{X}}(k + 1 | k) = \mathbf{A}X(k) + \mathbf{B}U(k) + \mathbf{H}\omega(k)$
    comment: Compute the cost function. Equation 25
    $J = \| \hat{e}_{i\alpha s}(k + 1 | k) + \hat{e}_{i\beta s}(k + 1 | k) \|^2$
    if $J < J_o$ **then**
      $J_o \leftarrow J, S^{opt} \leftarrow \mathbf{S}_i$
    end if
    $i := i + 1$
  end while
  comment: Compute the correction of the covariance matrix. Equation 24
  $\varphi(k + 1) = \mathbf{A} \cdot \Gamma(k) \cdot \mathbf{A}^T + \mathbf{H} \cdot R_\omega \cdot \mathbf{H}^T$

---

4. Simulation results

A Matlab/Simulink simulation environment has been designed for the VSI-fed asynchronous asymmetrical dual three-phase induction machine, and simulations have been done to prove the efficiency of the scheme proposed. Numerical integration using fourth order Runge-Kutta

---

[3] 12 active, corresponding to the largest vectors in the $(\alpha - \beta)$ subspace and the smallest ones in the $(x - y)$ subspace plus a zero vector.

algorithm has been applied to compute the evolution of the state variables step by step in the time domain. Table 3 shows the electrical and mechanical parameters for the asymmetrical dual three-phase induction machine.

Parameter		Value
Stator resistance	R_s (Ω)	1.63
Rotor resistance	R_r (Ω)	1.08
Stator inductance	L_s (H)	0.2792
Rotor inductance	L_r (H)	0.2886
Mutual inductance	L_m (H)	0.2602
Inertia	J_i (kg.m^2)	0.109
Pairs of poles	P	3
Friction coefficient	B (kg.m^2/s)	0.021
Nominal frequency	w_a (Hz)	50

Table 3. Parameters of the asymmetrical dual three-phase induction machine

Computer simulations allow valuing the effectiveness of the proposed control system under unload and full-load conditions, with respect to the mean squared error (MSE) of the speed and stator current tracking. In all cases is considered a sampling frequency of 6.5 kHz, and that the initial conditions of the covariance matrix ($\varphi(0)$), and the process and measurement noise, are known. The Kalman Filter has been started with the following initial conditions; $\varphi(0) = diag \begin{bmatrix} 1 & 1 & 1 & 1 \end{bmatrix}$, in order to indicate that the initial uncertainty (rms) of the state variables is 1 A. Because $\varphi(k)$ is time varying, the KF gain is sensitive to this initial condition estimate during the initial transient, but the steady final values are not affected [11]. The magnitudes of the process noise (R_ω) and measurement noise (R_V) are known and are generate using a Random Source block of the Simulink Signal Processing Blockset, assuming the following values, $R_\omega = 15 \times 10^{-3}$ and $R_V = 25 \times 10^{-3}$, respectively.

4.1. Efficiency of current control loop

A series of simulation tests are performed in order to verify the efficiency of current control loop in three points of operation of the machine. Figure 5 shows the current tracking in stationary reference frame ($\alpha - \beta$) and ($x - y$) subspaces considering sub-optimal solution in the optimization process (12 active and 1 null vectors). The predicted stator current in the α component is shown in the upper side (zoom graphs and green curves). For all cases of analysis efficiency is measured with respect to the MSE of the currents tracking in ($\alpha - \beta$)-($x - y$) subspaces and the total harmonic distortion (THD), defined as the ratio of the sum of the powers of all harmonic components to the power of the fundamental frequency, obtained from the Powergui-Continuous Simulink block. A 2.5 A reference stator current at 12 Hz is established for the case of Figure 5 (a). Figure 5 (b) shows the current tracking in the ($\alpha - \beta$) and ($x - y$) subspaces using a 2 A reference stator current at 18 Hz and Figure 5 (c) shows the current tracking in stationary reference frame using a 1.5 A reference stator current at 36 Hz. Table 4 summarizes the results of the three previous trials where are considered different amplitudes and angular frequencies for the reference current.

From the obtained results can be emphasized as follows:

Method	Test	MSE_α, MSE_β	MSE_x, MSE_y	THD_α, THD_β
	(a)	0.2105, 0.2322	0.9298, 0.9304	7.1330, 7.5969
MBPC	(b)	0.1989, 0.2141	1.0957, 1.0885	10.3610, 11.8192
	(c)	0.2287, 0.2348	1.2266, 1.3102	15.8951, 17.4362

Table 4. Simulation results obtained from Figure 5

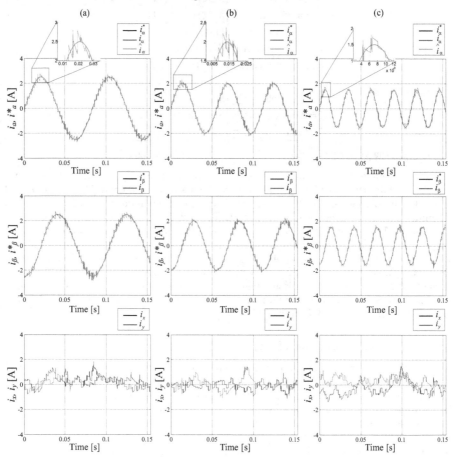

Figure 5. Stator current in $(\alpha - \beta)$ component tracking and $(x - y)$ current components. (a) 2.5 A (peak) current reference at 12 Hz. (b) 2 A current reference at 18 Hz and (c) 1.5 A current reference at 36 Hz

a. The MBPC is a flexible approach that, opposite to PWM based control methods, allows a straightforward generalization to different requirements only changing the cost function

b. The MBPC method is discontinuous technique, so the switching frequency is unknown. This feature reduces the switching losses (compared to continuous techniques) at expense of an increase in the harmonics of the stator current

c. As increases the frequency of the reference currents the switching frequency decreases, consequently there is a degradation in the THD of the stator currents as can be seen in Table 4

4.2. Proposed speed control method

The structure of the proposed speed control for the asymmetrical dual three-phase induction machine based on a KF is shown in Figure 6. The process of calculation of the slip frequency (ω_{sl}) is performed in the same manner as the Indirect Field Orientation methods, from the reference currents in dynamic reference frame (i_{ds}^*, i_{qs}^*) and the electrical parameters of the machine (R_r, L_r) [13, 14]. The inner loop of the current control, based on the MBPC selects control actions solving an optimization problem at each sampling period using a real system model to predict the outputs. As the rotor current can not be measured directly, it should be estimated using a reduced order estimator based on an optimal recursive estimation algorithm from the Equations 21-24.

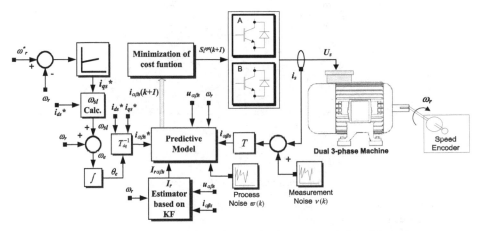

Figure 6. Proposed speed control technique based on KF for the asymmetrical dual three-phase induction machine

Different cost functions (J) can be used, to express different control criteria. The absolute current error, in stationary reference frame ($\alpha - \beta$) for the next sampling instant is normally used for computational simplicity. In this case, the cost function is defined as Equation 25, where i_s^* is the stator reference current and $i_{\alpha\beta}(k+1|k)$ is the predicted stator current which is computationally obtained using the predictive model. However, other cost functions can be established, including harmonics minimization, switching stress or VSI losses [6]. Proportional integral (PI) controller is used in the speed control loop, based on the indirect vector control schema because of its simplicity. In the indirect vector control scheme, PI speed controller is used to generate the reference current i_{ds}^* in dynamic reference frame. The current reference used by the predictive model is obtained from the calculation of the electric angle used to convert the current reference, originally in dynamic reference frame ($d - q$), to static reference frame ($\alpha - \beta$) as shown in Figure 6.

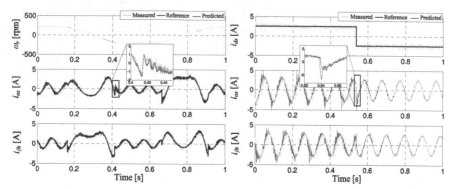

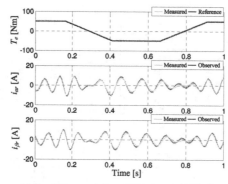

(a) Simulation results for a ±320 rpm step wave speed (b) Simulation results for a ±2.5 A step in the reference
comand tracking   current comand tracking

(c) Simulation results for a 50 Nm trapezoidal load

Figure 7. Simulation results for a proposed speed control. The predicted stator current in the α
component is shown in the upper side (zoom graphs, red curves)

Figure 7 (a) shows simulation results for a 200 revolutions per minute (rpm) trapezoidal
speed reference, if we consider a fixed current reference ($i^*_{ds} = 1$ A). The subscripts $(\alpha - \beta)$
represent quantities in the stationary frame reference of the stator currents. The measurement
speed is fed back in the closed loop for speed regulation and a PI controller is used in the
speed regulation loop as shown in Figure 6. The predicted stator current in the α component
is shown in the upper side (zoom graph, red curve). Under these test conditions, the
MSE in the speed and current tracking are 0.75 rpm and 0.15 A, respectively. Figure 7 (b)
shows the step response for the induction machine to a change of ±2.5 A in the current
reference (i^*_{ds} see Figure 6), if we consider a fixed speed reference ($\omega^*_r = 200$ rpm). In these
simulation results, the subscripts $(\alpha - \beta)$ represent the stator current in stationary reference
frame. Under these test conditions, the MSE in the stator current tracking are 0.1 A for the
reference current (i^*_{ds}) and 0.18 A considering stationary reference frame. Finally, Figure 7 (c)
shows a trapezoidal load application response, and the rotor current evolution (measured and
observed) in stationary reference frame. These simulation results substantiate the expected

performance of the proposed algorithm, based on a Kalman Filter. The estimated rotor current converges to real values for these test conditions as shown in figures, proving that the observer performance is satisfactory.

5. Conclusions

In this chapter a new approach for the speed control of the asymmetrical dual three-phase induction machine has been proposed and evaluated. The speed control scheme uses an inner loop predictive current control based on the model, where the main advantage is the absence of modulation techniques. The MBPC is described using a state-space representation, where the rotor and stator current are the states variables. The proposed algorithm provides an optimal estimation of the rotor current in each sampling time in a recursive manner, even when internal parameters of the drive are not precisely known, and the measurements of the state variables are perturbed by gaussian noise. The theoretical development based on a Kalman Filter has been validated by simulations results. The method has proven to be efficient even when considering that the machine is operating under varying load regimes.

Acknowledgments

The author gratefully the Paraguay Government for the economical support provided by means of a Conacyt Grant (project 10INV01). Also, wishes to express his gratitude to the anonymous reviewers for their helpful comments and suggestions.

Author details

Raúl Igmar Gregor Recalde
Engineering Faculty of the National University of Asunción
Department of Power and Control Systems, Asunción-Paraguay

6. References

[1] Levi, E. (2008). Multiphase electric machines for variable-speed applications. *IEEE Transactions on Industrial Electronics*, Vol. 55, No. 5, (May 2008) page numbers (1893-1909), ISSN 0278-0046

[2] Bucknal R. & Ciaramella, K. (2010). On the Conceptual Design and Performance of a Matrix Converter for Marine Electric Propulsion. *IEEE Transactions on Power Electronics*, Vol. 25, No. 6, (June 2010) page numbers (1497-1508), ISSN 0885-8993

[3] Zhao Y. & Lipo, T. (1995). Space vector PWM control of dual three-phase induction machine using vector space decomposition. *IEEE Transactions on Industry Applications*, Vol. 31, No. 5, (October 1995) page numbers (1100-1109), ISSN 0093-9994

[4] Boglietti, A.; Bojoi, R.; Cavagnino, A.& Tenconi, A. (2008). Efficiency Analysis of PWM Inverter Fed Three-Phase and Dual Three-Phase High Frequency Induction Machines for Low/Medium Power Applications. *IEEE Transactions on Industrial Electronics*, Vol. 55, No. 5, (May 2008) page numbers (2015-2023), ISSN 0278-0046

[5] Arahal, M.; Barrero, F.; Toral, S.; Durán, M.; & Gregor, R. (2008). Multi-phase current control using finite-state model-predictive control. *Control Engineering Practice*, Vol. 17, No. 5, (October 2008) page numbers (579-587), ISSN 0967-0661

[6] Barrero, F.; Arahal, M.; Gregor, R.; Toral, S. & Durán, M. (2009). A proof of concept study of predictive current control for VSI driven asymmetrical dual three-phase AC machines. *IEEE Transactions on Industrial Electronics*, Vol. 56, No. 6, (June 2009) page numbers (1937-1954), ISSN 0278-0046

[7] Barrero, F.; Prieto, J.; Levi, E.; Gregor, R.; Toral, S.; Duran, M. & Jones, M. (2011). An Enhanced Predictive Current Control Method for Asymmetrical Six-phase Motor Drive. *IEEE Transactions on Industrial Electronics*, Vol. 58, No. 8, (Aug. 2011) page numbers (3242-3252), ISSN 0278-0046

[8] Gregor, R.; Barrero, F.; Toral, S.; Duran, M.; Arahal, M.; Prieto, J. & Mora, J. (2010). Predictive-space vector PWM current control method for asymmetrical dual three-phase induction motor drives. *IET Electric Power Applications*, Vol. 4, No. 1, (January 2010) page numbers (26-34), ISSN 1751-8660

[9] Duran, M.; Prieto, J.; Barrero, F. & Toral, S. (2011). Predictive Current Control of Dual Three-Phase Drives Using Restrained Search Techniques. *IEEE Transactions on Industrial Electronics*, Vol. 58, No. 8, (Aug. 2011) page numbers (3253-3263), ISSN 0278-0046

[10] Shi, K.L.; Chan, T.F.; Wong, Y.K. & Ho, S.L. (2002). Speed estimation of an induction motor drive using an optimized extended Kalman filter. *IEEE Transactions on Industrial Electronics*, Vol. 49, No. 1, (Feb. 2002) page numbers (124-133), ISSN 0278-0046

[11] Franklin, G.; Powell, J. & Workman, M. (1998). Optimal Estimation. The Kalman Filter, In: *Digital Control of Dynamic Systems*, Addison - Wesley, (Ed. 3rd), page numbers (389-392), ISBN 978-0-201-82054-6

[12] Vargas, R.; Cortes, P.; Ammann, U.; Rodriguez, J. & Pontt, J. (2007). Predictive Control of a Three-Phase Neutral-Point-Clamped Inverter. *Transactions on Industrial Electronics*, Vol. 54, No. 5,(Oct. 2007), page numbers (2697-2705), ISSN 0278-0046

[13] Ong, C.M. (1997). Indirect Field Orientation Methods, In: *Dynamic Simulation of Electric Machinery Using MatLab/Simulink*, Prentice Hall, page numbers (439-440), ISBN 978-0-137-23785-2

[14] Krause, P.; Wasynczuk, O. & Sudhoff, S. (2002). Indirect Rotor Field-Oriented Control, In: *Analysis of Electric Machinery and Drive Systems*, Wiley-IEEE Press, (Ed. 2nd) page numbers (550-554), ISBN 978-0-471-14326-0

Chapter 7

Sensorless Control of Induction Motor Supplied by Current Source Inverter

Marcin Morawiec

Additional information is available at the end of the chapter

1. Introduction

The circuits used for power conversion applied in drives with induction motor (IM) are classified into two groups: voltage source inverters (VSI) and current source inverters (CSI). The VSI were used more often than the CSI because of their better properties. Nowadays, the development of power electronics devices has enormous influence on applications of systems based on the CSI and creates new possibilities.

In the 1980s the current source inverters were the main commonly used electric machine feeding devices. Characteristic features of those drives were the motor electromagnetic torque pulsations, the voltage and current with large content of higher harmonics. The current source inverter was constructed of a thyristor bridge and large inductance and large commutation capacitors. Serious problems in such drive systems were unavoidable overvoltage cases during the thyristor commutation, as the current source inverter current is supplied in a cycle from a dc-link circuit to the machine phase winding. The thyristor CSI has been replaced by the transistor reverse blocking IGBT devices (RBIGBT), where the diode is series-connected and placed in one casing with transistor. The power transistors like RBIGBT or Silicon Carbide (SiC) used in the modern CSIs guarantee superior static and dynamic drive characteristics.

The electric drive development trends are focused on the high quality system. The use of current sources for the electric machine control ensures better drive properties than in case of voltage sources, where it may be necessary to use an additional passive filter at the inverter output. The Pulse width modulation (PWM) with properly chosen dc-link inductor and input-output capacitors result in sinusoidal inverter output currents and voltages. Methods of calculating proper inductance in dc-link were proposed in [Glab (Morawiec) M. et. al., 2005, Klonne A. & Fuchs W.F., 2003, 2004]. Properties of dc-link circuit of the Current Source Converter (CSC) force the utilization of two fully-controlled inverters to supply system with electric motor. The first of them – CSI - generates the current output vector to supply the induction motor. The second one – Current Source Rectifier (CSR) - generates a DC voltage to supply a dc-link circuit. The strategy for controlling the output current vector of CSI can be realized in two ways [Glab (Morawiec) M. et. al., 2005, Klonne A. & Fuchs W.F., 2003, Kwak S. & Toliyat H.A., 2006]. First of them is based on changes of modulation index while the value of current in dc-link circuit remains constant [Klonne A. & Fuchs W.F., 2003]. The second method is based on changes of dc-link current. In this case the CSI is working with constant, maximum value of PWM modulation index. Control of modulation index in CSI is used in drive systems, where high dynamic of electromagnetic torque should be maintained [Klonne A. & Fuchs W.F., 2003]. High current in dc-link circuit is a reason for high power losses in CSI. The simplified control method is the scalar control: current to slip (I/s). This method is very simple to implement, but the drive system has average performance (only one controller is necessary, the current in dc-link is kept at constant value by PI controller).

The drive system quality is closely connected with the machine control algorithm. The space vector concept, introduced in 1959 by Kovacs and Racz, opened a new path in the electric machine mathematical modelling field. The international literature on the subject presents drive systems with the CSI feeding an induction motor with the control system based on the coordinate system orientation in relation to the rotor flux vector (FOC – Field Oriented Control). Such control consisted of the dc-link circuit current stabilization [Klonne A. & Fuchs W.F., 2003]. In such control systems the control variables are the inverter output current components. This control method is presented in [Nikolic Aleksandar B. & Jeftenic Borislav I.], where the authors analyze control system based on direct torque control. The control process where the control variable is the inverter output current may be called *current control* of an induction motor supplied by the CSI.

Another control method of a current source inverter fed induction motor is using the link circuit voltage and the motor slip as control variables. That type of control may be called *voltage control* of a CSI fed induction motor, as the dc-link circuit voltage and angular frequency of current vector are the control variables. Proposed control strategy bases on nonlinear multi-scalar control [Glab (Morawiec) M. et. al., 2005,]. The nonlinear control may result in better properties in case if the IM is fed by CSI. To achieve independent control of flux and rotor speed, new nonlinear control scheme is proposed. In this control method the inverter output currents are not controlled variables. The voltage in dc-link and pulsation of output current vector are the controlled variables which can be obtained by nonlinear transformations and are proposed by authors in [Krzeminski Z., 1987, Glab (Morawiec) M. et. al., 2005, 2007]. The multi-scalar model is named the extended one because the mathematical model contained dc-link current and output capacitors equations. This full mathematical model of induction machine with the CSI is used to derive new multi-scalar model. In proposed method the output current vector coefficients are not controlled variables. The output current vector and the flux vector are used to achieve new multi-scalar variables and new multi-scalar model. The control system structure may be supported on PI controllers and nonlinear decouplings or different controllers e.g. sliding mode controllers, the backstepping control method or fuzzy neural controller.

2. Structure of the drive system supplied by CSI or CSC

The simplified configuration of the drive system with the CSI is presented in the Fig. 1. The integral parts of the system are the inductor in dc-link and the output capacitors. In the Fig. 1 the structure with the chopper as an adjustable voltage source is presented.

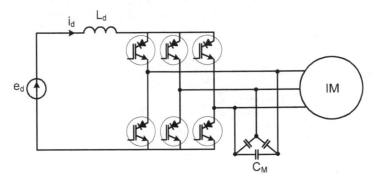

Figure 1. The CSI with the chopper

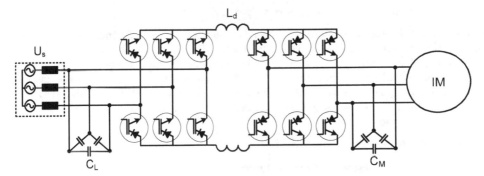

Figure 2. The Current source converter

The chopper with the small inductor L$_d$ (a few mH) forms the large dynamic impedance of the current source. In the proposed system the transistors forms commutator which transforms DC current into AC current with constant modulation index. The current is controlled by voltage source e$_d$ in dc-link. In this way the system with CSI remains voltage controlled and the differential equation for dc-link may be integrated with differential equation for the stator. The inductor limits current ripples during commutations of transistors. The transistors used in this structure are named the reverse blocking IGBT transistors (RBIGBT).

In order to avoid resonance problem the CSI or CSC structure parameters (input-output capacitors and inductor) ought to be properly chosen. The transistor CSI or CSC structures should guarantee sinusoidal stator current and voltage of IM if the parameters are selected by iteration algorithm.

2.1. The iteration algorithm selection of the inductor and input-output capacitors

The inductance in dc-link L_d may be calculated as a function of the integral versus time of the difference of input voltage e_d and output voltage u_d in dc-link [Glab (Morawiec) M. et. al., 2005, Klonne A. & Fuchs W.F., 2003, 2004]. Calculating an inductance from [Glab (Morawiec) M. et. al., 2005, Klonne A. & Fuchs W.F., 2003] may be not enough because of the resonance problem. The parameters could be determined by simple algorithm.

Two criteria are taken into account:

- Minimization of currents ripples in the system
- Minimization of size and weight.

The first criteria can be defined as:

$$w_i = \frac{\Delta i_{d\max}}{i_d},$$

(1)

where

Δi_{dmax} is $\max(i_{dmax}(t_1) - i_{dmin}(t_2))$,

i_d is average value of dc-link current in one period.

The current ripples in dc-link has influence on output currents and commutation process. According to this, the w_i factor, THD_i stator and THD_u stator must be taken into account. Optimal value of inductance L_d and output C_M ensure performance of the drive system with sinusoidal output current and small THD. In Fig. 3 the iteration algorithm for choosing the inductor and capacitor is shown. In every step of iteration new values w_i, THD_i, THD_u are received. In every of these steps new values are compared with predetermined value w_{ip}, THD_{ip}, THD_{up} and:

N is number of iteration,

THD_i – stator current total harmonic distortion,

THD_u – stator voltage total harmonic distortion. Number of iteration is set for user.

In START the initial parameters are loaded. In block Set L_d inductance of the inductor is set. In Numerical process block the simulation is started. In next steps THD_i, THD_u and w_i coefficient are calculated. THD_i, THD_u and w_i coefficient are compared with predetermined value. If YES then C_M is setting, if NO the new value of L_d must be set. Comparison with predetermined value is specified as below:

$$\begin{cases} THD_{ip\max} \leq THD_i(i) \leq THD_{ip\min} \\ THD_{up\max} \leq THD_u(i) \leq THD_{up\min} \\ \Delta i_{dp\max} \leq \Delta i_d(i) \leq \Delta i_{dp\min} \end{cases}$$

(2)

where

Δ_{idpmax} – maximum value for w_i coefficient,

Δ_{idpmin} – minimum value for w_i coefficient,

THD_{ipmax}, THD_{upmax} – maximum predetermined value of THD for range (ΔL_d, ΔC_M),

THD_{ipmin}, THD_{upmin} – minimum predetermined value of THD (ΔL_d, ΔC_M),

ΔL_d – interval of optimal value L_d,

ΔC_M – interval of optimal value C_M.

For optimal quality of stator current and voltage in a drive system THD_i ought to be about 1%, $THD_u < 2\%$ and $w_i < 15\%$ in numerical process. Estimated CSC parameters by the iteration algorithm are shown in Fig. 4 and 5 or Table 1.

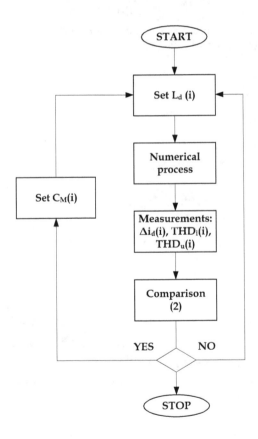

Figure 3. The iteration algorithm for selection of the inductor L_d an capacitors C_M

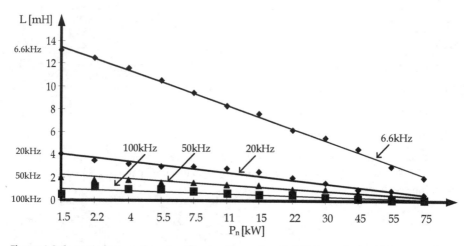

Figure 4. Inductor inductance from iteration algorithm, where P_n [kW] is nominal machine power for different transistors switching frequency [kHz]

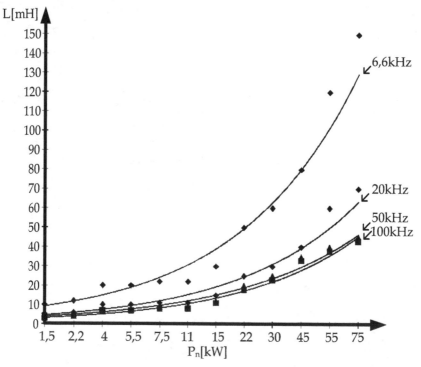

Figure 5. Capacitor capacitance from iteration algorithm where P_n [kW] is nominal machine power for different transistors switching frequency [kHz]

The value of AC side capacitors C_L ought to be about 25% higher than for C_M because of higher harmonics in supply network voltage:

$$C_L \approx 1,25 \cdot C_M \,. \tag{3}$$

P_n [kW]	L_d [mH]	C_M [µF]	C_L [µF]	P_n [kW]	L_d [mH]	C_M [µF]	C_L [µF]
1,5	13,2	10	10	15	7,6	30	35
2,2	12,5	12	12	22	6,2	50	60
4	11,6	20	20	30	5,5	60	70
5,5	10,5	20	20	45	4,5	80	90
7,5	9,4	22	22	55	3	120	150
11	8,3	22	25	75	2	150	200

Table 1. Estimated a CSC parameters

3. The mathematical model of IM supplied by CSC

3.1. Introduction to mathematical model

Differential equation for the dc-link is as follows

$$e_d = i_d R_d + L_d \frac{di_d}{d\tau} + u_d \,, \tag{4}$$

where: u_d is the inverter input voltage, R_d is the inductor resistance, L_d is the inductance, e_d is the control voltage in dc-link, i_d is the current in dc-link.

Equation (4) is used together with differential equation for the induction motor to derive the models of induction motor fed by the CSI.

The model of a squirrel-cage induction motor expressed as a set of differential equations for the stator-current and rotor-flux vector components presented in $\alpha\beta$ stationary coordinate system is as follows [Krzeminski Z., 1987]:

$$\frac{di_{s\alpha}}{d\tau} = -\frac{R_s L_r^2 + R_r L_m^2}{L_r w_\sigma} i_{s\alpha} + \frac{R_r L_m}{L_r w_\sigma} \psi_{r\alpha} + \omega_r \frac{L_m}{w_\sigma} \psi_{r\beta} + \frac{L_r}{w_\sigma} u_{s\alpha} \,, \tag{5}$$

$$\frac{di_{s\beta}}{d\tau} = -\frac{R_s L_r^2 + R_r L_m^2}{L_r w_\sigma} i_{s\beta} + \frac{R_r L_m}{L_r w_\sigma} \psi_{r\beta} - \omega_r \frac{L_m}{w_\sigma} \psi_{r\alpha} + \frac{L_r}{w_\sigma} u_{s\beta} \,, \tag{6}$$

$$\frac{d\psi_{r\alpha}}{d\tau} = -\frac{R_r}{L_r} \psi_{r\alpha} - \omega_r \psi_{r\beta} + \frac{R_r L_m}{L_r} i_{s\alpha} \,, \tag{7}$$

$$\frac{d\psi_{r\beta}}{d\tau} = -\frac{R_r}{L_r} \psi_{r\beta} + \omega_r \psi_{r\alpha} + \frac{R_r L_m}{L_r} i_{s\beta} \,, \tag{8}$$

$$\frac{d\omega_r}{d\tau} = \frac{L_m}{JL_r}(\psi_{r\alpha}i_{s\beta} - \psi_{r\beta}i_{s\alpha}) - \frac{1}{J}m_0,$$ (9)

where

R_r, R_s are the motor windings resistance, L_s, L_r, L_m are stator, rotor and mutual inductance, $u_{s\alpha}$, $u_{s\beta}$, $i_{s\alpha}$, $i_{s\beta}$, $\psi_{r\alpha}$, $\psi_{r\beta}$ are components of stator voltage, currents and rotor flux vectors, ω_r is the angular rotor velocity, J is the torque of inertia, m_0 is the load torque. All variables and parameters are in p. u.

3.2. The mathematical model of IM contains full drive system equations

The vector components of the rotor flux together with inverter output current are used to derive model of IM fed by the CSI. The model is developed using rotating reference frame xy with x axis orientated with output current vector. The y component of the output current vector is equal to zero.

The variables in the rotating reference frame are presented in Fig. 6.

The output current under assumption an ideal commutator can be expressed

$$i_f = K \cdot i_d,$$ (10)

where

K is the unitary commutation function (K=1).

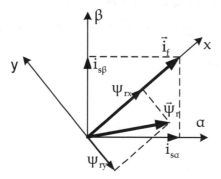

Figure 6. Variables in the rotating frame of references

If the commutation function is K=1 than

$$|i_f| \approx i_d.$$ (11)

The equation (12) results from (11) taking into account ideal commutator of the CSI, according to equation

$$p_{DClink} = p_{AC\ motor\ side} ,$$
$$u_d i_d = u_{sx} i_{fx} \qquad (12)$$

where: u_d is the input six transistors bridge voltage, u_{sx} is the stator voltage component.

The full model of the drive system in rotating reference frame xy with x axis oriented with inverter output current vector is as follows

$$\frac{di_{sx}}{d\tau} = -\frac{R_s L_r^2 + R_r L_m^2}{L_r w_\sigma} i_{sx} + \frac{R_r L_m}{L_r w_\sigma} \psi_{rx} + \omega_i i_{sy} + \omega_r \frac{L_m}{w_\sigma} \psi_{ry} + \frac{L_r}{w_\sigma} u_{sx}, \qquad (13)$$

$$\frac{di_{sy}}{d\tau} = -\frac{R_s L_r^2 + R_r L_m^2}{L_r w_\sigma} i_{sy} + \frac{R_r L_m}{L_r w_\sigma} \psi_{ry} - \omega_i i_{sx} - \omega_r \frac{L_m}{w_\sigma} \psi_{rx} + \frac{L_r}{w_\sigma} u_{sy}, \qquad (14)$$

$$\frac{d\psi_{rx}}{d\tau} = -\frac{R_r}{L_r} \psi_{rx} + (\omega_i - \omega_r)\psi_{ry} + \frac{R_r L_m}{L_r} i_{sx}, \qquad (15)$$

$$\frac{d\psi_{ry}}{d\tau} = -\frac{R_r}{L_r} \psi_{ry} - (\omega_i - \omega_r)\psi_{rx} + \frac{R_r L_m}{L_r} i_{sy}, \qquad (16)$$

$$\frac{d\omega_r}{d\tau} = \frac{L_m}{JL_r} (\psi_{rx} i_{sy} - \psi_{ry} i_{sx}) - \frac{1}{J} m_0, \qquad (17)$$

$$\frac{di_d}{d\tau} = \frac{e_d}{L_d} - \frac{R_d}{L_d} i_d - \frac{u_{sx}}{L_d}, \qquad (18)$$

$$\frac{du_{sx}}{d\tau} = \frac{1}{C_M} (i_{fx} - i_{sx}) + \omega_i u_{sy}, \qquad (19)$$

$$\frac{du_{sy}}{d\tau} = -\frac{1}{C_M} i_{sy} - \omega_i u_{sx}. \qquad (20)$$

where: ω_i is angular frequency of vector $\vec{i}_f$, i_{sx}, i_{sy} are the capacitors currents.

4. The nonlinear multi-scalar voltage control of IM with PI controllers

4.1. The simplified Multi-scalar control

The Nonlinear multi-scalar control was presented by authors [Krzeminski Z., 1987, Glab (Morawiec) M. et. al., 2005, 2007]. This control in classical form based on PI controllers. The simplify multi-scalar control of IM supplied by CSC for different vector components ($\vec{\psi}_r, \vec{i}_s$), ($\vec{\psi}_s, \vec{i}_s$), ($\vec{\psi}_m, \vec{i}_s$) was presented in [Glab (Morawiec) M. et. al., 2005, 2007]. These multi-scalar control structures give different dynamical and statical properties of IM supplied by CSI. In this chapter the simplified control is presented. The simplification is based on (11) and (12) equations. If the capacity C_M has small values (a few μF) the mathematical equations (19) - (20) can be ommitted and the output current vector in stationary state is $|i_f| \approx |i_s|$. Under this simplification, to achieve the decoupling between two control paths the multi-scalar model based control system was proposed [Krzeminski Z., 1987, Glab (Morawiec) M. et. al., 2005, 2007]. The variables for the multi-scalar model of IM are defined

$$x_{11} = \omega_r, \tag{21}$$

$$x_{12} = -i_d \psi_{ry}, \tag{22}$$

$$x_{21} = \psi_{rx}^2 + \psi_{ry}^2, \tag{23}$$

$$x_{22} = i_d \psi_{rx}, \tag{24}$$

where

x_{11} is the rotor speed, x_{12} is the variable proportional to electromagnetic torque, x_{21} is the square of rotor flux and x_{22} is the variable named magnetized variable [Krzeminski, 1987].

Assumption of such machine state variables may lead to improvement of the control system quality due to the fact that e.g. the x_{12} variable is directly the electromagnetic torque of the machine. In FOC control methods [Klonne A. & Fuchs W.F., 2003, 2004, Salo M. & Tuusa H. 2004] the electromagnetic torque is not directly but indirectly controlled (the i_{sq} stator current component). With the assumption of a constant rotor flux modulus, such a control conception is correct. The inaccuracy of the machine parameters, asymmetry or inadequately aligned control system may lead to couplings between control circuits.

The mathematical model for new state of variables (21) - (24) used (15) - (18) is expressed by differential equations:

$$\frac{dx_{11}}{d\tau} = \frac{L_m}{JL_r}(x_{12}) - \frac{1}{J}m_0, \tag{25}$$

$$\frac{dx_{12}}{d\tau} = -\frac{1}{T_i}x_{12} + \frac{1}{L_d}u_{sx}\psi_{ry} - \frac{R_r L_m}{L_r}i_{sy}i_d + v_1, \tag{26}$$

$$\frac{dx_{21}}{d\tau} = -2\frac{R_r}{L_r}x_{21} + 2R_r\frac{L_m}{L_r}x_{22}, \tag{27}$$

$$\frac{dx_{22}}{d\tau} = -\frac{1}{T_i}x_{22} + \frac{R_r L_m}{L_r}i_{sx}i_d + v_2. \tag{28}$$

The compensation of nonlinearities in differential equation leads to the following expressions for control variables v_1 and v_2 appearing in differential equations (27) - (28):

$$v_1 = \frac{R_r L_m}{L_r} i_{sy} i_d - \frac{u_{sx}}{L_d} \cdot \psi_{ry} + \frac{1}{T_i} m_1, \tag{29}$$

$$v_2 = -\frac{R_r L_m}{L_r} i_{sx} i_d + \frac{u_{sx}}{L_d} \cdot \psi_{rx} + \frac{1}{T_i} m_2, \tag{30}$$

where $m_{1,2}$ are the PI controllers output and

$$\frac{1}{T_i} = \frac{R_s}{L_r} + \frac{R_d}{L_d}. \tag{31}$$

The control variables are specified

$$e_d = -L_d \cdot \frac{\psi_{ry} v_1 - \psi_{rx} v_2}{x_{21}}, \tag{32}$$

$$\omega_i = \frac{\psi_{rx} v_1 + \psi_{ry} v_2}{i_d \cdot x_{21}} + x_{11}, \tag{33}$$

when $x_{21}, i_d \neq 0$.

The inverter control variables are: voltage e_d and the output current vector pulsation. The multi-scalar control of IM supplied by CSI was named voltage control because the control variable is voltage e_d in dc-link.

The decoupled two subsystems are obtained:

• electromagnetic subsystem

$$\frac{dx_{21}}{d\tau} = -2\frac{R_r}{L_r} x_{21} + 2\frac{R_r L_m}{L_r} x_{22}, \tag{34}$$

$$\frac{dx_{22}}{d\tau} = \frac{1}{T_i}(-x_{22} + m_2), \tag{35}$$

• electromechanical subsystem

$$\frac{dx_{11}}{d\tau} = \frac{L_m}{JL_r} x_{12} - \frac{1}{J} m_0, \tag{36}$$

$$\frac{dx_{12}}{d\tau} = \frac{1}{T_i}(-x_{12} + m_1). \tag{37}$$

4.2. The multi-scalar control with inverter mathematical model

The author in [Morawiec M., 2007] revealed stability proof of simplified multi-scalar control while the parameters of the CSI are optimal selected.

When the capacitance C_M is neglected the stator current vector $\vec{i}_s$ is about ~5% out of phase to $\vec{i}_f$ while nominal torque is set. Then the control variables and decoupling are not obtained precisely. The error is small than 2% because PI controllers improved it.

In order to compensate these errors the capacity C_M to mathematical model is applied.

From (19) - (20) in stationary state lead to dependences:

$$i_{sx} = i_{fx} + \omega_{if} C_M u_{sy} ,\tag{38}$$

$$i_{sy} = -\omega_{if} C_M u_{sx} .\tag{39}$$

The new mathematical model of the drive system is obtained from (38) - (39) through differentiation it and used (15) - (16) in xy coordinate system:

$$\frac{di_{sx}}{d\tau} = -\frac{R_d}{L_d}i_d + \frac{1}{L_d}e_d - \frac{1}{L_d}u_x - \omega_{if}C_M i_{sy} - \omega_{if}^2 C_M^2 u_{sx} ,\tag{40}$$

$$\frac{di_{sy}}{d\tau} = -\omega_{if}C_M i_d + \omega_{if}C_M i_{sx} - \omega_{if}^2 C_M^2 u_{sy} ,\tag{41}$$

$$\frac{d\psi_{rx}}{d\tau} = -\frac{R_r}{L_r}\psi_{rx} + (\omega_{if} - \omega_r)\psi_{ry} + \frac{R_r L_m}{L_r}i_{sx} ,\tag{42}$$

$$\frac{d\psi_{ry}}{d\tau} = -\frac{R_r}{L_r}\psi_{ry} - (\omega_{if} - \omega_r)\psi_{rx} + \frac{R_r L_m}{L_r}i_{sy} ,\tag{43}$$

$$\frac{di_d}{d\tau} = -\frac{R_d}{L_d}i_d + \frac{1}{L_d}e_d - \frac{1}{L_d}u_{sx} ,\tag{44}$$

$$\frac{du_{sx}}{d\tau} = \frac{1}{C_M}(i_{fx} - i_{sx}) + \omega_{if}u_{sy} ,\tag{45}$$

$$\frac{du_{sy}}{d\tau} = -\frac{1}{C_M}i_{sy} - \omega_{if}u_{sx} .\tag{46}$$

Substituting (38) - (39) to multi-scalar variables [Krzeminski Z., 1987] one obtains:

$$x_{11} = \omega_r ,\tag{47}$$

$$x_{12} = -i_{fx}\psi_{ry} - \omega_{if}C_M x_{32} , \tag{48}$$

$$x_{21} = \psi_{rx}^2 + \psi_{ry}^2 , \tag{49}$$

$$x_{22} = i_{fx}\psi_{rx} + \omega_{if}C_M x_{31} , \tag{50}$$

and

$$x_{31} = \psi_{rx}u_{sy} - \psi_{ry}u_{sx} , \tag{51}$$

$$x_{32} = \psi_{rx}u_{sx} + \psi_{ry}u_{sy} . \tag{52}$$

The multi-scalar model for new multi-scalar variables has the form:

$$\frac{dx_{11}}{d\tau} = \frac{L_m}{JL_r}x_{12} - \frac{1}{J}m_0 , \tag{53}$$

$$\frac{dx_{12}}{d\tau} = -\frac{1}{T_i}x_{12} + \frac{1}{L_d}u_{sx}\psi_{ry} - x_{11}x_{22} + v_1 , \tag{54}$$

$$\frac{dx_{21}}{d\tau} = -2\frac{R_r}{L_r}x_{21} + 2R_r\frac{L_m}{L_r}x_{22} , \tag{55}$$

$$\frac{dx_{22}}{d\tau} = -\frac{1}{T_i}x_{22} - \frac{1}{L_d}u_{sx}\psi_{rx} + \frac{R_r L_m}{L_r}i_d^2 + x_{11}x_{12} + v_2 , \tag{56}$$

where

$$v_1 = -\frac{1}{L_d}e_d\psi_{ry} + \omega_{if}(x_{22} - \frac{R_d}{L}C_M x_{32} - C_M \frac{R_r L_m}{L_r}p_s) , \tag{57}$$

$$v_2 = \frac{1}{L_d}e_d\psi_{rx} + \omega_{if}(-x_{12} + \frac{R_d}{L}C_M x_{31} + C_M \frac{R_r L_m}{L_r}q_s) , \tag{58}$$

$$q_s = i_{sx}u_{sy} - i_{sy}u_{sx} , \tag{59}$$

$$p_s = u_{sx}i_{sx} + u_{sy}i_{sy} . \tag{60}$$

The compensation of nonlinearities in differentials equation leads to the following expressions for control variables v_1 and v_2 appearing in differential equations (54), (56):

$$v_1 = \frac{1}{T_i}m_1 - \frac{1}{L_d}u_{sx}\psi_{ry} + x_{11}x_{22} , \tag{61}$$

$$v_2 = \frac{1}{T_i} m_2 + \frac{1}{L_d} u_{sx} \psi_{rx} - x_{11} x_{12} - \frac{R_r L_m}{L_r} i_d^2 , \tag{62}$$

and the control variables

$$e_d = L_d \frac{V_2 x_{41} - V_1 x_{42}}{\psi_{rx} x_{41} + \psi_{ry} x_{42}} , \tag{63}$$

$$\omega_i = \frac{V_1 \psi_{rx} + V_2 \psi_{ry}}{\psi_{rx} x_{41} + \psi_{ry} x_{42}} , \tag{64}$$

where

$$x_{41} = x_{22} - \frac{R_d}{L} C_M x_{32} - C_M \frac{R_r L_m}{L_r} p_s , \tag{65}$$

$$x_{42} = -x_{12} + \frac{R_d}{L} C_M x_{31} + C_M \frac{R_r L_m}{L_r} q_s , \tag{66}$$

$\frac{1}{T_i}$ is determined in (31).

The decoupled two subsystems are obtained as in (34) - (37).

4.3. The multi-scalar adaptive-backstepping control of an IM supplied by the CSI

The backstepping control can be appropriately written for an induction squirrel-cage machine supplied from a VSI. In literature the backstepping control is known for adaptation of selected machine parameters, written for an induction motor [Tan H. & Chang J., 1999, Young Ho Hwang, 2008]. In [Tan H. & Chang J., 1999, Young Ho Hwang, 2008] the authors defined the machine state variables in the dq coordinate system, oriented in accordance with the rotor flux vector (FOC). The control method presented in [Tan H. & Chang J., 1999, Young Ho Hwang, 2008] is based on control of the motor state variables: ω_r – rotor angular speed, rotor flux modulus and the stator current vector components: i_{sd} and i_{sq}. Selection of the new motor state variables, as in the case of multi-scalar control with linear PI regulators, leads to a different form of expressions describing the machine control and decoupling. The following state variables have been selected for the multi-scalar backstepping control

$$e_1 = x_{11}^* - x_{11} , \tag{67}$$

$$e_2 = x_{12}^* - x_{12} , \tag{68}$$

$$e_3 = x_{21}^* - x_{21} , \tag{69}$$

$$e_4 = \left(2R_r \frac{L_m}{L_r} x_{22}\right)^* - 2R_r \frac{L_m}{L_r} x_{22} , \tag{70}$$

where: x_{11}, x_{12}, x_{21} and x_{22} are defined in (47) - (50).

The e_4 tracking error is defined in (70), it does not influence on the control system properties and is only an accepted simplification in the format of decoupling variables.

Derivatives of the (67) - (70) errors take the form

$$\dot{e}_1 = \frac{L_m}{JL_r} e_2 - k_1 e_1 - \frac{\tilde{m}_0}{J} , \tag{71}$$

$$\dot{e}_2 = k_1 e_2 - k_1^2 \frac{JL_r}{L_m} e_1 + \frac{L_r}{L_m} \dot{\tilde{m}}_0 + \frac{1}{T_i} x_{12} - \frac{1}{L_d} u_{sx} \psi_{ry} + x_{11} x_{22} + \frac{R_r L_m}{L_r} i_{sy} i_d - v_1 , \tag{72}$$

$$\dot{e}_3 = -k_3 e_3 + e_4 , \tag{73}$$

$$\dot{e}_4 = -k_3^2 e_3 + k_3 e_4 - 4\left(\frac{R_r}{L_r}\right)^2 x_{21} + 4\left(\frac{R_r}{L_r}\right)^2 L_m x_{22} + 2\frac{R_r L_m}{L_r T_i} x_{22} + 2\frac{R_r L_m}{L_r L_d} u_{sx} \psi_{rx} - 2\left(\frac{R_r L_m}{L_r}\right)^2 i_{sx} i_d +$$
$$-2\frac{R_r L_m}{L_r} x_{11} x_{12} - 2\frac{R_r L_m}{L_r} v_2 \tag{74}$$

The Lyapunov function derivative, with (71) – (74) taken into account, may be expressed as:

$$\dot{V} = -k_1^2 e_1 - k_2^2 e_2 - k_3^2 e_3 - k_4^2 e_4 + e_2(f_1 - v_1) + e_4(f_2 - a_3 v_2) +$$
$$+ \tilde{m}_0 \left(-\frac{e_1}{J} - k_1 \frac{L_r}{L_m} e_2 + \frac{\dot{\tilde{m}}_0}{\gamma}\right) \tag{75}$$

where

$$v_1 = -\frac{1}{L_d} e_d \psi_{ry} + \omega_{if} x_{41} , \tag{76}$$

$$v_2 = \frac{1}{L_d} e_d \psi_{rx} + \omega_{if} x_{42} , \tag{77}$$

$$f_1 = \lim it_{12} \cdot e_1 \left(\frac{L_m}{JL_r} - k_1^2 \frac{JL_r}{L_m}\right) + k_2 e_2 + k_1 e_2 + \frac{L_r}{L_m} \dot{\tilde{m}}_0 + \frac{1}{T_i} x_{12} - \frac{1}{L_d} u_{sx} \psi_{ry} +$$
$$+ x_{11} x_{22} \tag{78}$$

$$f_2 = \lim it_{12} \cdot (e_3 - k_3^2 e_3) + k_4 e_4 + k_3 e_4 - 4\left(\frac{R_r}{L_r}\right)^2 x_{21} + 4\left(\frac{R_r}{L_r}\right)^2 L_m x_{22} + 2\frac{R_r L_m}{L_r T_i} x_{22} +$$
$$+ 2\frac{R_r L_m}{L_r L_d} u_{sx} \psi_{rx} - 2\left(\frac{R_r L_m}{L_r}\right)^2 i_d^2 - 2\frac{R_r L_m}{L_r} x_{11} x_{12} \tag{79}$$

$$a_3 = 2\frac{R_r L_m}{L_r}.$$

limit$_{12}$ – is a dynamic limitation in the motor speed control subsystem,

limit$_{22}$ – is a dynamic limitation in the rotor flux control subsystem,

$k_1 \ldots k_4$ and γ are the constant gains.

The control variables take the form:

$$e_d = L_d \frac{x_{41} f_2 - a_3 x_{42} f_1}{a_3 (\psi_{rx} x_{41} + \psi_{ry} x_{42})}, \tag{80}$$

$$\omega_i = \frac{a_3 \psi_{rx} f_1 + \psi_{ry} f_2}{a_3 (\psi_{rx} x_{41} + \psi_{ry} x_{42})}. \tag{81}$$

The inverter control variables are: voltage e_d and the output current vector pulsation. The two decoupled subsystems are obtained as in (34) - (37).

The load torque m_0 can be estimated from the formula:

$$\dot{\hat{m}}_0 = \gamma(\frac{e_1}{J} + k_1 \frac{L_r}{L_m} e_2). \tag{82}$$

4.4. Dynamic limitations of the reference variables

In control systems with the conventional linear controllers of the PI or PID type, the reference (or controller output) variable dynamics are limited to a constant value or dynamically changed by (83) - (84), depending on the drive working point.

Control systems where the control variables are determined from the Lyapunov function (like in backstepping control) have no limitations in the set variable control circuits. The reference variable dynamics may be limited by means of additional first order inertia elements (e.g. on the set speed signal).

The author of this paper has not come across a solution of the problem in the most significant backstepping control literature references, e.g. [Tan H. & Chang J., 1999, Young Ho Hwang, 2008]. In the quoted reference positions, the authors propose the use of an inertia elements on the set variable signals. Such approach is an intermediate method, not giving any rational control effects. The use of an inertia element on the reference signal, e.g. of the rotor angular speed, will slow down the reference electromagnetic torque reaction in proportion to the inertia element time-constant. In effect a "slow" build-up of the motor electromagnetic torque is obtained, which may be acceptable in some applications. In practice the aim is to limit the electromagnetic torque value without an impact on the build-up dynamics. Control systems with the Lyapunov function-based control without limitation of the set variables are not suitable for direct adaptation in the drive systems. Therefore, a solution often quoted in literature is the use of a PI or PID speed controller at the torque control circuit input.

The set values of the x_{12}^*, x_{22}^* variables appearing in the e_2 and e_4 deviations can be dynamically limited and the dynamic limitations are defined by the expressions [Adamowicz M.; Guzinski J., 2005]:

$$x_{12\lim} = \sqrt{I_{s\max}^2 x_{21} - x_{22}^2} \, , \tag{83}$$

$$x_{22\lim} = f(U_{s\max}^2, I_{s\max}^2, x_{11}) \, , \tag{84}$$

where

$x_{12\lim}$ – the set torque limitation,

$x_{22\lim}$ – the x_{22} variable limitation,

$I_{s\max}$ – maximum value of the stator current modulus,

$U_{s\max}$ – maximum value of the stator voltage modulus.

The above given expressions may be modified to:

$$x_{12\lim} = \sqrt{I_{s\max}^2 x_{21} - \frac{x_{21}^2}{L_m^2}} \, , \tag{85}$$

giving the relationship between the x_{21} variable, the stator current modulus $I_{s\max}$, and the motor set torque limitation.

For the multi-scalar backstepping control, to the f_1 and f_2 variables the limit$_{12}$ and limit$_{22}$ variables were introduced; they assume the 0 or 1 value depending on the need of limiting the set variable.

Limitation of variables in the Lyapunov function-based control systems may be performed in the following way:

$$if \left(x_{12}^* > x_{12\lim} \right) then \begin{cases} \text{limit}_{12} = 0, \\ e_2 = x_{12\lim} - x_{12} \end{cases} , \tag{86}$$

$$if \left(x_{12}^* < -x_{12\lim} \right) then \begin{cases} \text{limit}_{12} = 0, \\ e_2 = -x_{12\lim} - x_{12} \end{cases} , \tag{87}$$

$$else \ \text{limit}_{12} = 1 \, ,$$

$$if \left(x_{22}^* > x_{22\lim} \right) then \begin{cases} \text{limit}_{22} = 0, \\ e_4 = x_{22\lim} - x_{22} \end{cases} , \tag{88}$$

$$if \left(x_{22}^{*} < -x_{22\,lim} \right) \ then \ \begin{cases} limit_{22} = 0, \\ e_4 = -x_{22\,lim} - x_{22} \end{cases}, \tag{89}$$

$$else \ limit_{22} = 1.$$

The dynamic limitations effected in accordance with expressions (83) – (84) limit properly the value of x_{12}^{*} and x_{22}^{*} variables without any interference in the reference signal build-up dynamics.

Fig. 7 presents the variable simulation diagrams. The backstepping control dynamic limitations were used.

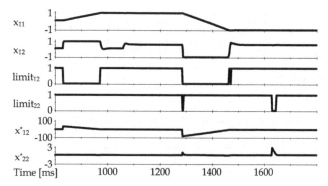

Figure 7. Diagrams of multi-scalar variables in the machine dynamic states, the $x_{12ogr} = 1.0$ and $x_{22ogr} = 0.74$ limitations were set for a drive system with an induction squirrel-cage machine supplied from a CSC-simulation diagrams, x^{*}_{12} – diagram of the machine set electromagnetic torque (without signal limitation), x^{*}_{22} – diagram of the x^{*}_{22} set signal (without limitation).

4.5. Impact of the dynamic limitation on the estimation of parameters

The use of a variable limitation algorithm may have a negative impact on the control system estimated parameters. This has a direct connected with the limited deviation values, which are then used in an adaptive parameter estimation. Such phenomenon is presented in Fig. 7. The estimated parameter in the control system is the motor load torque $\hat{m}_0$. The set electromagnetic torque is limited to the $x_{12lim} = 1.0$ value. Fig. 7 shows that the estimated load torque increases slowly in the intermediate states. Limitation of the set electromagnetic torque causes the limitation of deviation e_2, which in turn causes limited increase dynamics of the estimated load torque. The $\hat{m}_0$ value for $limit_{12} = 0$ in the dynamic states does not reach the real value of the load torque, which should be $\hat{m}_0 \approx x_{12}$. A large $\tilde{m}_0$ estimation error occurs in the intermediate states, which can be seen in Fig. 8. The estimation error in the intermediate states is $\tilde{m}_0 \neq 0$ because the torque limitation, introduced to the control system, is not compensated. The simulation and experimental tests have shown that the load torque estimation error in the intermediate state has an insignificant impact on the speed control. Omitting $\hat{m}_0$ in the set torque x^*_{12} expression eliminates the intermediate state speed over-regulation. But absence of $\hat{m}_0$ in x^*_{12} for a steady state gives the deviation value $e_1 \neq 0$ and lack of full control over maintaining the rotor set angular speed. Compensation of the $limit_{12}$ limitation introduced to the control system is possible by installing a corrector in the rotor angular speed control circuit.

A corrector in the form of an e_1 signal integrating element was added to the set electromagnetic torque x^*_{12} signal. In this way a system reacting to the change of machine real load torque was obtained. The introduced correction minimizes the rotor angular speed deviation and the corrector signal may be treated as the estimated load torque value.

The correction element is determined by the expression:

$$KT_L = k_{e1} \int_{t_{k-1}}^{t_k} e_1 d\tau ,$$ (90)

where

$t_{k-1}...t_k$ is the e_1 signal integration range,

KT_L – correction element,

k_{e1} – is the correction element amplification.

The gain k_{e1} should be adjusted that the speed overregulation in the intermediate state does not exceed 5%:

$$0 < k_{e1} \leq 0,1 \cdot k_1 ,$$ (91)

The correction element amplification must not be greater than k_1, or:

$$k_{e1} \leq k_1 .$$ (92)

For $k_{e1} > k_1$ the KT_L signal will become an oscillation element and may lead to the control system loss of stability.

The KT_L signal must be limited to the x_{12lim} value.

The x^*_{12} set value expression must be modified:

$$x^*_{12} = \frac{JL_r}{L_m} k_1 e_1 + KT_L ,$$ (93)

where

$$\hat{m}_0 \approx KT_L .$$ (94)

The use of (93) in the angular speed control circuit improves the load torque estimation and eliminates the steady state speed error.

Fig. 9 presents the load torque (determined in (94)) estimation as well as x_{12} and the limit$_{12}$ limitations.

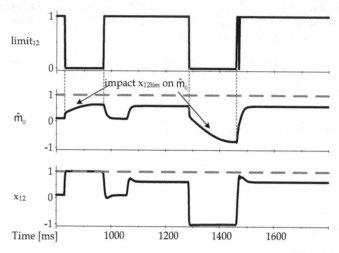

Figure 8. Impact of the electromagnetic torque limitation x_{12lim} on the estimated load torque $\hat{m}_0$ (82).

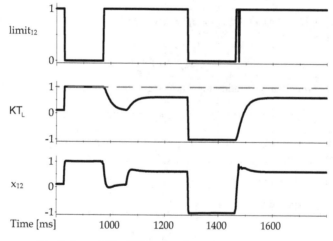

Figure 9. Diagrams of the limit$_{12}$ variable, KT_L load torque and electromagnetic torque x_{12}.

5. The nonlinear multi-scalar current control of induction machine

Conception of the CSI current control is based on forced components of the CSI output current. The dc-link circuit inductor could be modeled as the first order inertia element with the time constant T of a value equal to the dc-link circuit time constant value. The dc-link equation may be introduced to the induction machine mathematical model to obtain the set CSI output current component. The time constant T is equal the inductance L_d, it can be written:

$$\frac{d\vec{i}_s}{dt} = \frac{1}{T}(\vec{i}_f - \vec{i}_s),$$
(95)

$$T = \frac{R_d}{L_d}.$$
(96)

5.1. The simplified multi-scalar current control of induction machine

The simplified version of the CSI output current control it is assumed that the output capacitors have negligibly small capacitance, so their impact on the drive system dynamics is small. Assuming that the cartesian coordinate system, where the mathematical model variables are defined, is associated with the CSI output current vector (which with this simplification is the machine stator current) the mathematical model can be obtained (95) and (42) - (43) equations).

The multi-scalar variables have the form

$$x_{11} = \omega_r,$$
(97)

$$x_{12} = -i_{sx}\psi_{rx},$$
(98)

$$x_{21} = \psi_{rx}^2 + \psi_{ry}^2,$$
(99)

$$x_{22} = i_{sx}\psi_{rx},$$
(100)

where

i_{sx} is treated as the output current vector component and $|i_f| \approx |i_s|$.

For those variables, the multi-scalar model has the form:

$$\frac{dx_{11}}{d\tau} = \frac{L_m}{JL_r}x_{12} - \frac{1}{J}m_0,$$
(101)

$$\frac{dx_{12}}{d\tau} = -\frac{1}{T_i}x_{12} + v_1,$$
(102)

$$\frac{dx_{21}}{d\tau} = -2\frac{R_r}{L_r}x_{21} - 2\frac{R_r L_m}{L_r}x_{22},$$
(103)

$$\frac{dx_{22}}{d\tau} = -\frac{1}{T_i}x_{22} + \frac{R_r L_m}{L_r}i_d + v_2 \, . \tag{104}$$

Applying the linearization method, the following relations are obtained, where m_1 is the subordinated regulator output in the speed control line and m_2 is the subordinated regulator output in the flux control line

$$v_1 = \frac{1}{T_i}m_1 \, , \tag{105}$$

$$v_2 = \frac{1}{T_i}m_2 - \frac{R_r L_m}{L_r}i_d \, . \tag{106}$$

The control variables are modulus of the CSI output current and the output current vector pulsation, given by the following relations:

$$|i_f| = T\frac{v_2\psi_{rx} - v_1\psi_{ry}}{x_{21}\psi_{ry}} \, , \tag{107}$$

$$\omega_i = \frac{v_1 + v_2}{x_{21}i_{sx}} + x_{11} \, . \tag{108}$$

where: L_d –inductance, T_i– the system time constant.

5.2. The multi-scalar current control of induction machine

The current control analysis presented in the preceding sections does not take the CSI output capacitors into account. Such simplification may be applied because of the small impact of the capacitors upon the control variables (the machine stator current and voltage are measured). The capacitor model will have a positive impact on the control system dynamics.

The output capacitor relations have the form:

$$\frac{d\vec{u}_s}{dt} = \frac{1}{C_M}(\vec{i}_f - \vec{i}_s) \, , \tag{109}$$

where: $\vec{u}_s$ is the capacitor voltage vector, $\vec{i}_f$ is the current source inverter output current vector, $\vec{i}_s$ is the stator current vector.

Using the approximation method, relation (38) may be written as follows:

$$\frac{\vec{u}_s(k) - \vec{u}_s(k-1)}{T_{imp}} = \frac{1}{C_M}\left[\vec{i}_f(k) - \vec{i}_s(k)\right] \, . \tag{110}$$

Deriving $\mathbf{u}_s(k)$ from (38), the motor stator voltage is obtained as a function of the output current, stator current and stator voltage, in the form:

$$\vec{u}_s(k) = \frac{T_{imp}}{C_M}\left[\vec{i}_f(k) - \vec{i}_s(k)\right] + \vec{u}_s(k-1), \tag{111}$$

where

$\vec{u}_s(k)$ is the stator voltage vector at the k-th moment, T_{imp} - sampling period.

In the equations (5) - (9) representing the cage induction motor mathematical model the stator current vector components are appeared but the direct control variables do not. The motor stator current vector components cannot be the control variables because the multi-scalar model relations are derived from them. This is a different situation than with the FOC control. The FOC control is based on the machine stator current components described in a coordinate system associated with the rotor flux and the stator current components are the control variables. Therefore, control variables must be introduced into the mathematical model (5) - (9). The control may be introduced considering the machine currents (5) - (6) and equation (95) written for the $\alpha\beta$ components and describing the dc-link circuit dynamics. Adding the respective sides of equations (5) and (95) and equations (6) and (95), where equation (95) must be written with the $(\alpha\beta)$ components – the mathematical model of the drive system fed by the CSI is obtained:

$$\frac{di_{s\alpha}}{d\tau} = -\frac{R_s L_r^2 T + R_r L_m^2 T + L_r w_\sigma}{2L_r w_\sigma T}i_{s\alpha} + \frac{R_r L_m}{2L_r w_\sigma}\psi_{r\alpha} + \omega_r\frac{L_m}{2w_\sigma}\psi_{r\alpha} + \frac{L_r}{2w_\sigma}u_{s\alpha} + \frac{1}{2T}i_{f\alpha}, \tag{112}$$

$$\frac{di_{s\beta}}{d\tau} = -\frac{R_s L_r^2 T + R_r L_m^2 T + L_r w_\sigma}{2L_r w_\sigma T}i_{s\beta} + \frac{R_r L_m}{2L_r w_\sigma}\psi_{r\beta} - \omega_r\frac{L_m}{2w_\sigma}\psi_{r\alpha} + \frac{L_r}{2w_\sigma}u_{s\beta} + \frac{1}{2T}i_{f\beta}, \tag{113}$$

and equations (7) - (8).

The multi-scalar variables are assumed like in [Krzeminski Z., 1987]:

$$x_{11} = \omega_r, \tag{114}$$

$$x_{12} = \psi_{r\alpha}i_{s\beta} - \psi_{r\beta}i_{s\alpha}, \tag{115}$$

$$x_{21} = \psi_{r\alpha}^2 + \psi_{r\beta}^2, \tag{116}$$

$$x_{22} = \psi_{r\alpha}i_{s\alpha} + \psi_{r\beta}i_{s\beta}. \tag{117}$$

Introducing the multi-scalar variables (114) - (117), the multi-scalar model of an IM fed by the CSI is obtained:

$$\frac{dx_{12}}{d\tau} = -\frac{1}{2T_i}x_{12} - x_{11}x_{22} - x_{11}x_{22}\frac{L_m}{2w_\sigma} + \frac{L_r}{2w_\sigma}(u_{s\beta}\psi_{ra} - u_{sa}\psi_{r\beta}) + v_1, \qquad (118)$$

$$\frac{dx_{22}}{d\tau} = -\frac{1}{2T_i}x_{22} + x_{11}x_{12} + \frac{R_r L_m}{2L_r w_\sigma}x_{21} + \frac{R_r L_m}{2L_r}i_{sa}^2 + \frac{L_r}{2w_\sigma}(\psi_{ra}u_{sa} + \psi_{r\beta}u_{s\beta}) + v_2, \qquad (119)$$

where

$$v_1 = \frac{1}{2T}\psi_{ra}i_{f\beta} - \frac{1}{2T}\psi_{r\beta}i_{fa}, \qquad (120)$$

$$v_2 = \frac{1}{2T}\psi_{ra}i_{fa} + \frac{1}{2T}\psi_{r\beta}i_{f\beta}. \qquad (121)$$

Applying the linearization method to (118) - (119), the following expressions are obtained:

$$v_1 = \frac{1}{2T_i}m_1 + x_{11}x_{22} + \frac{L_m}{2w_\sigma}x_{11}x_{21} + \frac{L_r}{2w_\sigma}\left[u_{s\beta}(k-1)\psi_{ra} - u_{sa}(k-1)\psi_{r\beta}\right] - a_1 x_{12}, \qquad (122)$$

$$v_2 = \frac{1}{2T_i}m_2 - x_{11}x_{21} - \frac{R_r L_m}{2w_\sigma L_r}x_{21} - \frac{R_r L_m}{2L_r}i_{sa}^2 - \frac{L_r}{2w_\sigma}\left[u_{sa}(k-1)\psi_{ra} + u_{s\beta}(k-1)\psi_{r\beta}\right] + a_1 x_{22}, \qquad (123)$$

where

$$v_1 = a_1\left[i_{f\beta}\psi_{ra} - i_{fa}\psi_{r\beta}\right], \qquad (124)$$

$$v_2 = -a_1\left[i_{fa}\psi_{ra} + i_{f\beta}\psi_{r\beta}\right]. \qquad (125)$$

The control variables take the form

$$i_{fa} = a_2\frac{\psi_{ra}v_2 - \psi_{r\beta}v_1}{x_{21}}, \qquad (126)$$

$$i_{f\beta} = a_2\frac{\psi_{ra}v_1 + \psi_{r\beta}v_2}{x_{21}}, \qquad (127)$$

where

$$a_1 = \frac{L_r T_{imp}}{2w_\sigma C_M}, \qquad (128)$$

$$a_2 = a_1 + \frac{1}{2T}. \qquad (129)$$

The time constant T_i for simplified for both control method is given

$$T_i = \frac{w_\sigma L_r T}{R_s L_r^2 T + R_r L_m^2 T + L_r w_\sigma} ,$$ (130)

5.3. Generalized multi-scalar control of induction machine supplied by CSI or VSI

A cage induction machine fed by the CSI may be controlled in the same way as with the voltage source inverter (VSI). The generalized control is provided by an IM multi-scalar model formulated for the VSI machine control [Krzeminski Z., 1987]. The (114) - (117) multi-scalar variables and additional u_1 and u_2 variables are used

$$u_1 = \psi_{r\alpha} u_{s\beta} - \psi_{r\beta} u_{s\alpha} ,$$ (131)

$$u_2 = \psi_{r\alpha} u_{s\alpha} + \psi_{r\beta} u_{s\beta} ,$$ (132)

which are a scalar and vector product of the stator voltage and rotor flux vectors.

The multi-scalar model feedback linearization leads to defining the nonlinear decouplings [Krzeminski Z., 1987]:

$$U_1^* = \frac{w_\sigma}{L_r}\left[x_{11}(x_{22} + \frac{L_m}{w_\sigma}x_{21}) + \frac{1}{T_v}m_1 \right],$$ (133)

$$U_2^* = \frac{w_\sigma}{L_r}[-x_{11}x_{12} - \frac{R_r L_m}{L_r}i_s^2 - \frac{R_r L_m}{L_r w_\sigma}x_{21} + \frac{1}{T_v}m_2].$$ (134)

The control variables for an IM supplied by the VSI have the form [Krzeminski Z., 1987]:

$$u_{s\alpha}^* = \frac{\psi_{r\alpha} U_2^* - \psi_{r\beta} U_1^*}{x_{21}} ,$$ (135)

$$u_{s\beta}^* = \frac{\psi_{r\alpha} U_1^* + \psi_{r\beta} U_2^*}{x_{21}} .$$ (136)

The controls (135) - (136) are reference variables treated as input to space vector modulator when the IM is supplied by the VSI.

On the other side, when the IM is fed by the CSI, calculation of the derivatives of (131) - (132) multi-scalar variables yields the following relations:

$$\frac{du_1}{d\tau} = -\frac{R_r}{L_r}u_1 - x_{11}u_2 + \frac{R_r L_m}{L_r}q_s - \frac{1}{C_M}x_{12} + v_{11} ,$$ (137)

$$\frac{du_2}{d\tau} = -\frac{R_r}{L_r}u_2 + x_{11}u_1 + \frac{R_rL_m}{L_r}p_s + \frac{1}{C_M}x_{22} + v_{22}, \tag{138}$$

where p_s and q_s are defined in (59) - (60).

By feedback linearization of the system of equations, one obtains

$$v_{11} = -\frac{R_r}{L_r}v_{p1} - \frac{R_rL_m}{L_r}q_s + \frac{1}{C_M}x_{12} + x_{11}u_2, \tag{139}$$

$$v_{22} = -\frac{R_r}{L_r}v_{p2} - \frac{R_rL_m}{L_r}p_s + \frac{1}{C_M}x_{22} - x_{11}u_1, \tag{140}$$

where

v_{p1} and v_{p2} are the output of subordinated PI controllers.

The control variables of the IM fed by the CSI have the form:

$$i_{f\alpha} = -C_M\frac{v_{11}\psi_{r\beta} - v_{22}\psi_{r\alpha}}{x_{21}}, \tag{141}$$

$$i_{f\beta} = C_M\frac{v_{11}\psi_{r\alpha} + v_{22}\psi_{r\beta}}{x_{21}}. \tag{142}$$

As a result, two feedback loops and linear subsystems are obtained (Fig. 15).

6. The speed observer backstepping

General conception of the adaptive control with backstepping is presented in references [Payam A. F. & Dehkordi B. M. 2006, Krstic M.; Kanellakopoulos I.; & Kokotovic P. 1995]. In [Krstic M.; Kanellakopoulos I.; & Kokotovic P. 1995] the adaptive back integration observer stability is proved and the stability range is given.

Proceeding in accordance with the adaptive estimator with backstepping conception, one can derive formulae for the observer, where only the state variables will be estimated as well as the rotor angular speed as an additional estimation parameter.

Treating the stator current vector components $\hat{i}_{s\alpha,\beta}$ as the observer output variables (as in [Payam A. F. & Dehkordi B. M. 2006, Krstic M.; Kanellakopoulos I.; & Kokotovic P. 1995]) and $v_{\alpha,\beta}$ as the new input variables, which will be determined by the backstepping method, one obtains:

$$\frac{d\hat{i}_{s\alpha}}{d\tau} = -a_1\hat{i}_{s\alpha} + a_5\hat{\psi}_{r\alpha} + \hat{\omega}_r a_4\hat{\psi}_{r\beta} + a_6u_{s\alpha} + v_\alpha, \tag{143}$$

$$\frac{d\hat{i}_{s\beta}}{d\tau} = -a_1\hat{i}_{s\beta} + a_5\hat{\psi}_{r\beta} - \hat{\omega}_r a_4\hat{\psi}_{r\alpha} + a_6 u_{s\beta} + v_\beta \; , \tag{144}$$

$$\frac{d\hat{\psi}_{r\alpha}}{d\tau} = -\frac{R_r}{L_r}\hat{\psi}_{r\alpha} - \hat{\omega}_r\hat{\psi}_{r\beta} + \frac{R_r L_m}{L_r}\hat{i}_{s\alpha} \; , \tag{145}$$

$$\frac{d\hat{\psi}_{r\beta}}{d\tau} = -\frac{R_r}{L_r}\hat{\psi}_{r\beta} + \hat{\omega}_r\hat{\psi}_{r\alpha} + \frac{R_r L_m}{L_r}\hat{i}_{s\beta} \; . \tag{146}$$

In accordance to the backstepping method, the virtual control must be determined together with the observer stabilizing variables. In that purpose, the new ς_α and ς_β variables have been introduced and linked with the stator current estimation deviations (the integral backstepping structure [Krstic M.; Kanellakopoulos I.; & Kokotovic P. 1995]):

$$\frac{d\tilde{\varsigma}_a}{d\tau} = \tilde{i}_{s\alpha} \; , \tag{147}$$

$$\frac{d\tilde{\varsigma}_b}{d\tau} = \tilde{i}_{s\beta} \; . \tag{148}$$

The stator current vector component deviations are treated as the subsystem control variables [Payam A. F. & Dehkordi B. M. 2006, Krstic M.; Kanellakopoulos I.; & Kokotovic P. 1995]. Adding and deducting the stabilizing functions, one obtains:

$$\frac{d\tilde{\varsigma}_\alpha}{d\tau} = \tilde{i}_{s\alpha} - \sigma_\alpha + \sigma_\beta \; , \tag{149}$$

$$\frac{d\tilde{\varsigma}_\beta}{d\tau} = \tilde{i}_{s\beta} - \sigma_\alpha + \sigma_\beta \; , \tag{150}$$

where

$$\sigma_\alpha = -c_1\tilde{\varsigma}_\alpha \; , \quad \sigma_\beta = -c_1\tilde{\varsigma}_\beta \; , \tag{151}$$

by introducing the deviation defining variable, one obtains:

$$z_\alpha = \tilde{i}_{s\alpha} + c_1\tilde{\varsigma}_\alpha \; , \tag{152}$$

$$z_\beta = \tilde{i}_{s\beta} + c_1\tilde{\varsigma}_\beta \; . \tag{153}$$

Transformation of (152) - (153) leads to:

$$\frac{d\tilde{\varsigma}_\alpha}{d\tau} = z_\alpha - c_1\tilde{\varsigma}_\alpha \tag{154}$$

$$\frac{d\tilde{\varsigma}_\beta}{d\tau} = z_\beta - c_1\tilde{\varsigma}_\beta \ . \tag{155}$$

Calculation of the (152) - (153) deviation derivatives gives:

$$\dot{z}_\alpha = a_5\tilde{\psi}_{r\alpha} + a_4\left[\hat{\omega}_r\tilde{\psi}_{r\beta} + \tilde{\omega}_r(\hat{\psi}_{r\beta} - \tilde{\psi}_{r\beta})\right] + v_\alpha + c_1\tilde{i}_{s\alpha} \ , \tag{156}$$

$$\dot{z}_\beta = a_5\tilde{\psi}_{r\beta} - a_4\left[\hat{\omega}_r\tilde{\psi}_{r\alpha} + \tilde{\omega}_r(\hat{\psi}_{r\alpha} - \tilde{\psi}_{r\alpha})\right] + v_\beta + c_1\tilde{i}_{s\beta} \ . \tag{157}$$

By selecting the following Lyapunov function

$$V = \tilde{\varsigma}_\alpha^2 + \tilde{\varsigma}_\beta^2 + z_\alpha^2 + z_\beta^2 + \tilde{\psi}_{r\alpha}^2 + \tilde{\psi}_{r\beta}^2 + \frac{1}{\gamma}\tilde{\omega}_r^2 \ , \tag{158}$$

calculating the derivative and substituting the respective expressions, new correction elements can be determined, treated in the speed observer backstepping as the input variables. The Lyapunov function is determined for the dynamics of the $\varsigma_{\alpha,\beta}$, $z_{\alpha,\beta}$ variables and for the rotor flux components. Calculating the (158) derivative, one obtains:

$$\dot{V} = -c_1\tilde{\varsigma}_\alpha^2 - c_1\tilde{\varsigma}_\beta^2 - c_2z_\alpha^2 - c_2z_\beta^2 - \frac{R_r}{L_r}\tilde{\psi}_{r\alpha}^2 - \frac{R_r}{L_r}\tilde{\psi}_{r\beta}^2 + z_\alpha(a_5\tilde{\psi}_{r\alpha} + \hat{\omega}_r a_4\tilde{\psi}_{r\beta} + \tilde{\omega}_r a_4(\hat{\psi}_{r\beta} - \tilde{\psi}_{r\beta}) +$$
$$+v_\alpha + c_1\tilde{i}_{s\alpha} + c_2z_\alpha + \tilde{\varsigma}_\alpha) + z_\beta(a_5\tilde{\psi}_{r\beta} - \hat{\omega}_r a_4\tilde{\psi}_{r\alpha} - \tilde{\omega}_r a_4(\hat{\psi}_{r\alpha} - \tilde{\psi}_{r\alpha}) + v_\beta + c_1\tilde{i}_{s\beta} + c_2z_\beta + \tilde{\varsigma}_\beta) + \tag{159}$$
$$\tilde{\psi}_{r\alpha}(-\frac{R_r}{L_r}\tilde{\psi}_{r\alpha} - \hat{\omega}_r\tilde{\psi}_{r\beta} - \tilde{\omega}_r(\hat{\psi}_{r\beta} - \tilde{\psi}_{r\beta})) + \tilde{\psi}_{r\beta}(-\frac{R_r}{L_r}\tilde{\psi}_{r\beta} + \hat{\omega}_r\tilde{\psi}_{r\alpha} + \tilde{\omega}_r(\hat{\psi}_{r\alpha} - \tilde{\psi}_{r\alpha})).$$

The input variables $v_{\alpha,\beta}$, resulting directly from (159), should include the estimated variables and the estimation deviations:

$$v_\alpha = -a_5\tilde{\psi}_{r\alpha} - \hat{\omega}_r a_4\tilde{\psi}_{r\beta} - c_1\tilde{i}_{s\alpha} - c_2z_\alpha - \tilde{\varsigma}_\alpha \ , \tag{160}$$

$$v_\beta = -a_5\tilde{\psi}_{r\beta} + \hat{\omega}_r a_4\tilde{\psi}_{r\alpha} - c_1\tilde{i}_{s\beta} - c_2z_\beta - \tilde{\varsigma}_\beta \ . \tag{161}$$

Taking (160) - (161) into account, the deviation derivatives may be written in the form:

$$\dot{z}_\alpha = \tilde{\omega}_r a_4(\hat{\psi}_{r\beta} - \tilde{\psi}_{r\beta}) - c_2z_\alpha - \tilde{\varsigma}_\alpha \ , \tag{162}$$

$$\dot{z}_\beta = -\tilde{\omega}_r a_4(\hat{\psi}_{r\alpha} - \tilde{\psi}_{r\alpha}) - c_2z_\beta - \tilde{\varsigma}_\beta \ . \tag{163}$$

Using (162) - (163), the Lyapunov function may be written as follows:

$$\dot{V} = -c_1\tilde{\varsigma}_\alpha^2 - c_1\tilde{\varsigma}_\beta^2 - c_2z_\alpha^2 - c_2z_\beta^2 + \tilde{\omega}_r a_4\left[z_\alpha(\hat{\psi}_{r\beta} - \tilde{\psi}_{r\beta}) - z_\beta(\hat{\psi}_{r\alpha} - \tilde{\psi}_{r\alpha}) + \frac{1}{\gamma}\dot{\hat{\omega}}_r\right]. \tag{164}$$

The observer, defined by the (143) - (146) and (154) - (155) equations, is a backstepping type estimator.

In the (160) - (163) expressions the rotor flux deviations appear, which may be neglected without any change to the observer properties (143) - (146). Besides, the $\xi_{\alpha'\beta}$ deviations in (160) - (161) may be zero, thus lowering the observer order. Assuming the simplifications, one obtains

$$v_\alpha = -c_1 \tilde{i}_{s\alpha} - c_2 z_\alpha , \tag{165}$$

$$v_\beta = -c_1 \tilde{i}_{s\beta} - c_2 z_\beta , \tag{166}$$

and

$$\dot{\hat{\omega}}_r = \gamma a_4 \left(z_\beta \hat{\psi}_{r\alpha} - z_\alpha \hat{\psi}_{r\beta} \right). \tag{167}$$

where

c_1, c_2, γ are constant gains,

$$a_4 = \frac{L_m}{w_\sigma} , \; a_5 = \frac{R_r L_m}{L_r w_\sigma} , \; a_6 = \frac{L_r}{w_\sigma} .$$

In Fig. 10, 11 the backstepping speed observer test is shown. When the load torque is set to ~-0.1 p.u. the rotor speed in backstepping observer is more precisely estimated than e.g. Krzeminski's speed observer.

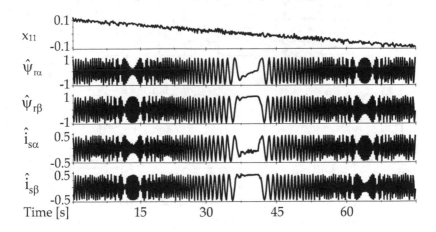

Figure 10. The Speed observer test: the estimated rotor speed x_{11} is changed from 0.1 to -0.1 p.u., the rotor flux and stator current coefficients are shown

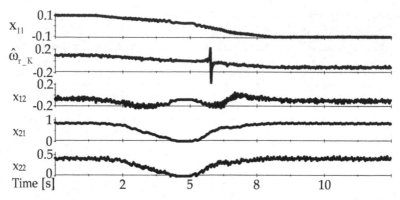

Figure 11. The Speed observer test: the estimated rotor speed x_{11} in backstepping observer is changed from 0.1 to -0.1 p.u., the estimated rotor speed $\hat{\omega}_{r\_K}$ by Krzeminski's speed observer [Krzeminski Z., 1999] and the multi-scalar variable: x_{12}, x_{21}, x_{22} are shown . The load torque m_0 is set to -0.1 p.u.

7. The control system structures

In Fig. 12 and Fig. 14 the voltage and current multi-scalar control system structure is shown. These structures are based on four PI controllers and contain: the modulator, the speed observer and decouplings blocks .

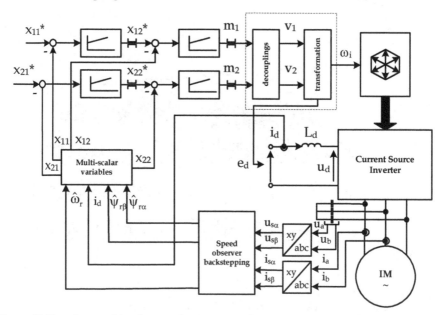

Figure 12. The voltage multi-scalar control system structure

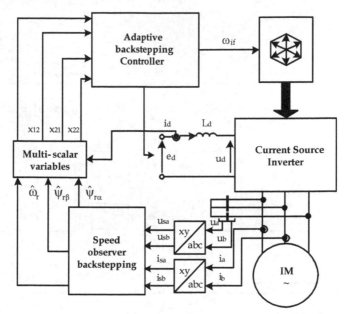

Figure 13. The voltage multi-scalar adaptive backstepping control system structure

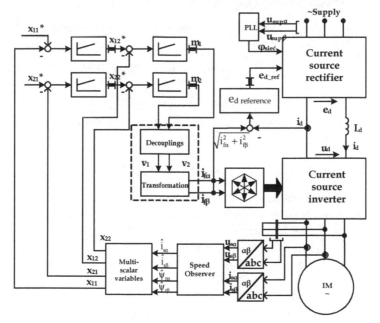

Figure 14. The current multi-scalar control system structure.

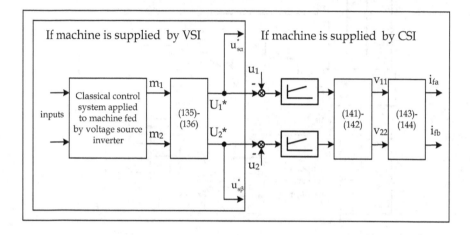

Figure 15. Generalized Multi-scalar Control System of Induction Machine supplied by CSI or VSI.

In Fig. 14 the e_{d_ref} value is determined in e_d reference block. The e_d reference block can be PI current controller or other controller.

In Fig. 13 the voltage multi-scalar adaptive backstepping control system structure is shown.

In Fig. 15 generalized multi-scalar control system structure is presented. This control structure is divided into two parts: the control system of IM fed by VSI and the control system of IM fed by CSI.

8. Experimental verification of the control systems

The tests were carried out in a 5.5 kW drive system. The motor parameters are given in Table 2 and the main per unit values in Table 3. In Fig. 16, 17 motor start-up and reverse for control system presented in chapter 4.1-4.2 are shown. In Fig. 18, 19 motor start-up and reverse for control system presented in chapter 5 are shown. Fig. 20,21 presents the same steady state like previous but for adaptive backstepping control system (chapter 4.3). Fig. 22 presents diagram of stator currents and voltages when motor is starting up for voltage control system (chapter 4.3). In Fig. 23 load torque setting to 0.7 p.u. for current control is presented. In Fig. 24, 25 the i_d current and the sinusoidal stator voltage and stator current are presented.

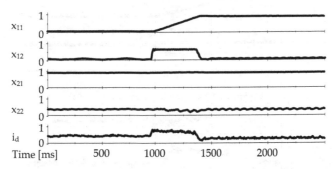

Figure 16. Motor start-up (chapter 4.1- 4.2)

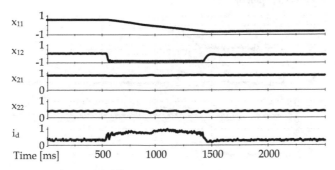

Figure 17. Motor reverse (chapter 4.1- 4.2)

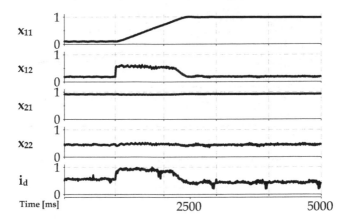

Figure 18. Motor start-up (chapter 5)

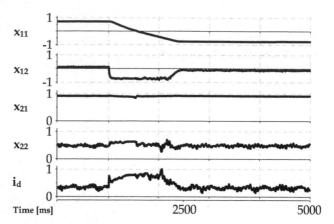

Figure 19. Motor reverse (chapter 5)

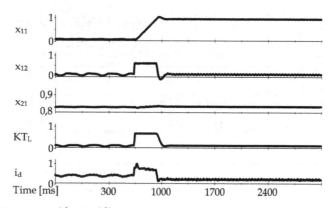

Figure 20. Motor start-up (chapter 4.3)

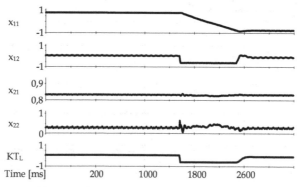

Figure 21. Motor reverse (chapter 4.3)

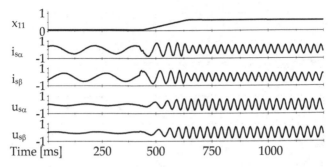

Figure 22. The currents and voltages

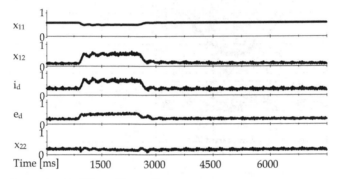

Figure 23. Load torque is set to 0.7 p.u. (chapter 5)

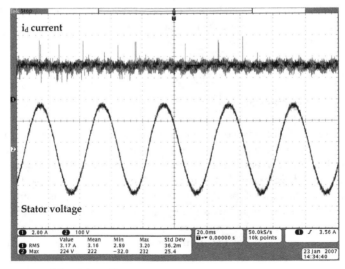

Figure 24. i_d current and the stator voltage

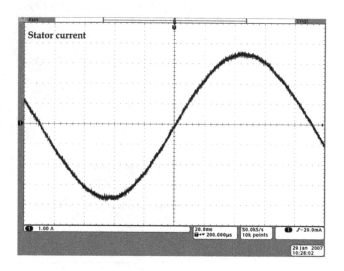

Figure 25. The stator current in stationary state.

where:

x_{11} is the rotor speed, x_{12} is the variable proportional to electromagnetic torque, x_{21} is the square of rotor flux and x_{22} is the additional variables, i_d is the dc-link current, $u_{s\alpha,\beta}$ are the capacitor voltage components, KTL is correction element (load torque), $i_{s\alpha,\beta}$ are the stator current components.

9. Conclusion

In this chapter two approaches to control of induction machine supplied by current source converter are presented. The first of them is voltage multi-scalar control based on PI controllers or backstepping controller. The voltage approach seems to be a better solution than the second one: current control, because the control system structure is more simple than the current control structure. The voltage in dc-link is the control variables obtained directly from decouplings. The current in dc-link is not kept at constant value but its value depend on induction machine working point. The current control gives higher losses in dc-link and higher transistor power losses than the voltage control. The power losses can be minimized by modulation index control method but the control system is more complicated. Both control systems lead to decoupling control path and sinusoidal stator current and voltage when space vector modulation of transistors is applied.

PARAMETER	VALUE
P_n (motor power)	5.5 kW
U_n (phase to phase voltage)	400 V
I_n (current)	10.9 A
J (interia)	0.0045 kgm²
n_n (rotor speed)	1500 rpm
PARAMETER	PER UNIT VALUES
R_s (stator resist.)	0.045
R_r (rotor resist.)	0.055
L_m (mutual-flux induct.)	1.95
L_s (stator induct.)	2.05
L_r (rotor induct.)	2.05
Current Source Converter	
C (capacitor in dc-link)	0.1
R_d (inductor resist.)	0.002
$C_{M,L}$ (input-output caps)	0.2

Table 2. The motor drive system parameters

DEFINITION	DESCRIPTION
$U_b = \sqrt{3}U_n$	base voltage
$I_b = I_n$	base current
$z_b = U_b / I_b$	base impedance

Table 3. Definition of per unit values

Author details

Marcin Morawiec
Gdansk University of Technology, Faculty of Electrical and Control Engineering, Poland

10. References

Adamowicz M.; Guzinski J.; Minimum-time minimum-loss speed sensorless control of induction motors under nonlinear control, *Compatibility in Power Electronics 2005*.

Bassi E.; Benzi F.P.; Bolognani S.; Buja G.S., A field orientation scheme for current-fed induction motor drives based on the torque angle closed-loop control, *IEEE Transactions on Industry Applications, Volume 28, Issue 5, Sept.-Oct. 1992 Pages: 1038 – 1044*.

Glab (Morawiec) M.; Krzeminski Z. & Włas M., The PWM current source inverter with IGBT transistors and multiscalar model control system, *11th European Conference on Power Electronics and Applications, IEEE 2005*.

Glab (Morawiec) M.; Krzeminski Z.; Lewicki A., Multiscalar control of induction machine supplied by current source inverter, *PCIM 2007, Nuremberg* 2007.

Fuschs F. & Kloenne A. dc-link and Dynamic Performance Features of PWM IGBT Current Source Converter Induction Machine Drives with Respect to Industrial Requirements, *IPEMC 2004, Vol. 3, 14-16 August 2004.*

Kwak S.; Toliyat H.A., A Current Source Inverter With Advanced External Circuit and Control Method, *IEEE Transactions on Industry Applications, Volume 42, Issue 6, Nov.-dec. 2006 Pages: 1496 – 1507.*

Klonne A. & Fuchs W.F., High dynamic performance of a PWM current source converter induction machine drive, *EPE 2003, Toulouse.*

Krstic M.; Kanellakopoulos I.; & Kokotovic P., Nonlinear and Adaptive Control Design, *John Wiley & Sons, 1995.*

Krzeminski Z., A new speed observer for control system of induction motor. IEEE Int. Conference on Power Electronics and Drive Systems, *PESC'99, Hong Kong, 1999.*

Krzeminski Z., Nonlinear control of induction motor, *Proceedings of the 10th IFAC World Congress, Munich 1987.*

Morawiec M., Sensorless control of induction machine supplied by current source inverter , *PhD Thesis, Gdansk University of Technology 2007.*

Nikolic Aleksandar B., Jeftenic Borislav I.: Improvements in Direct Torque Control of Induction Motor Supplied by CSI, *IEEE Industrial Electronics, IECON 2006 - 32nd Annual Conference on Industrial Electronics.*

Payam A. F.; Dehkordi B. M.; Nonlinear sliding-mode controller for sensorless speed control of DC servo motor using adaptive backstepping observer, *International Conference on Power Electr., PEDES '06, 2006.*

Salo M.; Tuusa H., Vector-controlled PWM current-source-inverter-fed induction motor drive with a new stator current control method, *IEEE Transactions on Industrial Electronics, Volume 52, Issue 2, April 2005 Pages: 523 – 531.*

Tan H.; & Chang J.; Adaptive Backstepping control of induction motor with uncertainties, *in Proc. the American control conference, California, June 1999, pp. 1-5.*

Young Ho Hwang; Ki Kwang Park; Hai Won Yang; Robust adaptive backstepping control for efficiency optimization of induction motors with uncertainties, *ISIE 2008.*

Chapter 8

Minimizing Torque-Ripple
in Inverter-Fed Induction Motor
Using Harmonic Elimination PWM Technique

Ouahid Bouchhida, Mohamed Seghir Boucherit and Abederrezzek Cherifi

Additional information is available at the end of the chapter

1. Introduction

Vector control has been widely used for the high-performance drive of the induction motor. As in DC motor, torque control of the induction motor is achieved by controlling torque and flux components independently. Vector control techniques can be separated into two categories: direct and indirect flux vector orientation control schemes. For direct control methods, the flux vector is obtained by using stator terminal quantities, while indirect methods use the machine slip frequency to achieve field orientation.

The overall performance of field-oriented-controlled induction motor drive systems is directly related to the performance of current control. Therefore, decoupling the control scheme is required by compensation of the coupling effect between q-axis and d-axis current dynamics (Jung et al., 1999; Lin et al., 2000; Suwankawin et al., 2002).

The PWM is the interface between the control block of the electrical drive and its associated electrical motor (fig.1). This function controls the voltage or the current inverter (VSI or CSI) of the drive. The performance of the system is influenced by the PWM that becomes therefore an essential element of the system. A few problems of our days concerning the variable speed system are related to the conventional PWM: inverter switching losses, acoustical noise, and voltages harmonics (fig.2).

Harmonic elimination and control in inverter applications have been researched since the early 1960's (Bouchhida et al., 2007, 2008; Czarkowski et al., 2002; García et al., 2003; Meghriche et al., 2004, 2005; Villarreal-Ortiz et al., 2005; Wells et al., 2004). The majority of these papers consider the harmonic elimination problem in the context of either a balanced connected load or a single phase inverter application. Typically, many papers have focused on finding solutions and have given little attention to which solution is optimal in an application context.

A pre-calculated PWM approach has been developed to minimize the harmonic ratio within the inverter output voltage (Bouchhida et al., 2007, 2008; Bouchhida, 2008, 2011). Several other techniques were proposed in order to reduce harmonic currents and voltages. Some benefit of harmonic reduction is a decrease of eddy currents and hysterisis losses. That increase of the life span of the machine winding insulation. The proposed approach is integrated within different control strategies of induction machine.

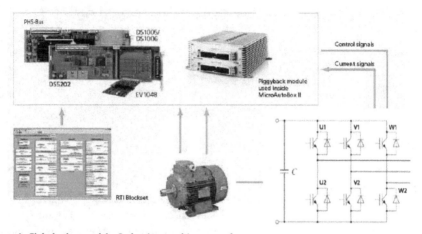

Figure 1. Global scheme of the Induction machine control

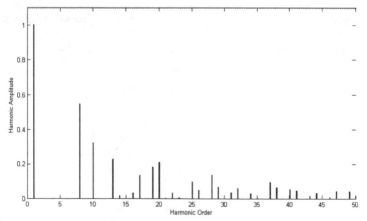

Figure 2. Harmonic spectrum of the PWM inverter output voltage.

A novel harmonic elimination pulse width modulated (PWM) strategy for three-phase inverter is presented in this chapter. The torque ripple of the induction motor can be significantly reduced by the new PWM technique. The three-phase inverter is associated with a passive LC filter. The commutation angles are predetermined off-line and stored in the microcontroller memory in order to speed up the online control of the induction motors. Pre-calculated switching is modelled to cancel the greater part of low-order harmonics and to keep a single-pole DC voltage across the polarized capacitors. A passive LC filter is designed to cancel the high-order harmonics. This approach allows substantial reduction of the harmonic ratio in the AC main output voltage without increasing the number of switches per period. Consequently, the duties of the semiconductor power switches are alleviated. The effectiveness of the new harmonic elimination PWM technique for reducing torque-ripple in inverter-fed induction motors is confirmed by simulation results. To show the validity of our approach, DSP-based experimental results are presented.

2. New three-phase inverter model

Figure 3 shows the new structure of the three-phase inverter, with E being the dc input voltage and $U_{12\text{-out}} = V_{C1} - V_{C2}$, $U_{23\text{-out}} = V_{C2} - V_{C3}$ and $U_{31\text{-out}} = V_{C3} - V_{C1}$ are the ac output voltage obtained via a three LC filter. R represents the internal inductors resistance. Q_i and Q'_i (i=1,2,3) are the semiconductor switches. It is worth to mention that transistors Q_i and Q'_i undergo complementary switching states. V_{C1}, V_{C2} and V_{C3} are the inverter filtered output voltages taken across capacitors C_1, C_2 and C_3 respectively.

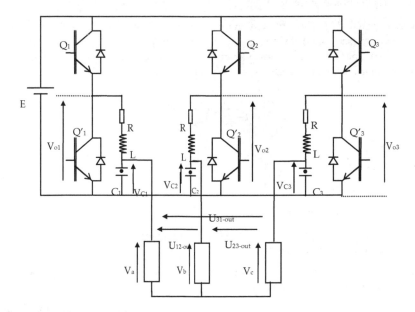

Figure 3. New three-phase inverter model.

2.1. Harmonic analysis

In ideal case, the non filtered three inverter output voltage V_{o1}, V_{o2} and V_{o3} is desired to be:

$$\begin{cases} V_{o1-ideal} = \dfrac{E}{2}[1 + \cos\alpha] \\[2mm] V_{o2-ideal} = \dfrac{E}{2}[1 + \cos(\alpha - \dfrac{2}{3}\pi)] \\[2mm] V_{o3-ideal} = \dfrac{E}{2}[1 + \cos(\alpha - \dfrac{4}{3}\pi)] \end{cases} \tag{1}$$

With: $\alpha = \omega t$, and ω is the angular frequency.

The relative Fourier harmonic coefficients of (1), with respect to E, are given by (2.1) or more explicitly by (2.2).

$$d_k^i = \frac{1}{E}\frac{1}{\pi}\int_{-\pi}^{+\pi} V_{oi-ideal} \cdot \cos(k\alpha) d\alpha \tag{2.1}$$

$$d_0^i = 1, \ d_1^i = \frac{1}{2}, \ d_k^i = 0 \text{, for } k \in \left[2, \infty\right[\tag{2.2}$$

With index i is the phase number.

However, in practice, the non filtered inverter output voltage V_{o1} (V_{o2}, V_{o3}) is a series of positive impulses (see Fig. 4): 0 when Q'_1 (Q'_2, Q'_3) is on and E when Q'_1(Q'_2, Q'_3) is off, so that voltage of capacitor C_1 (C_2, C_3) is always a null or a positive value. In this case, the relative Fourier harmonic coefficients, with respect to E, are given by (3).

$$a_k^1 = \frac{2}{k\pi}\sum_{i=0}^{N_\alpha} \sin k\alpha_i (-1)^{i+1} \tag{3}$$

Where:

k is the harmonic order

α_i are the switching angles

N_α is the number of α_i per half period

The other inverter outputs V_{o2} and V_{o3} are obtained by phase shifting V_{o1} with 2/3 π, 4/3 π, respectively as illustrated in fig. 4 for N_α=5.

The objective is to determine the switching angles α_i so as to obtain the best possible match between the inverter output V_{o1} and $V_{o1\text{-ideal}}$.

For this purpose, we have to compare their respective harmonics. A perfect matching is achieved only when an infinite number of harmonics is considered as given by (4).

$$\frac{2}{k\pi}\sum_{i=0}^{N_\alpha} sin\, k\alpha_i(-1)^{i+1} = d_k, \text{ for } k \in [0,\infty[\tag{4}$$

In practice, the number of harmonics N that can be identical is limited. Thus, a nonlinear system of N+1 equations having N_α unknowns is obtained as:

$$a_k = \frac{2}{k\pi}\sum_{i=0}^{N_\alpha} sin\, k\alpha_i(-1)^{i+1} = d_k, \text{ for } k \in [0,N] \tag{5}$$

To solve the nonlinear system (5), we propose to use the genetic algorithms, to determine the switching angles α_i (Bäck, 1996; Davis, 1991). The optimal switching angles family are listed in table I.

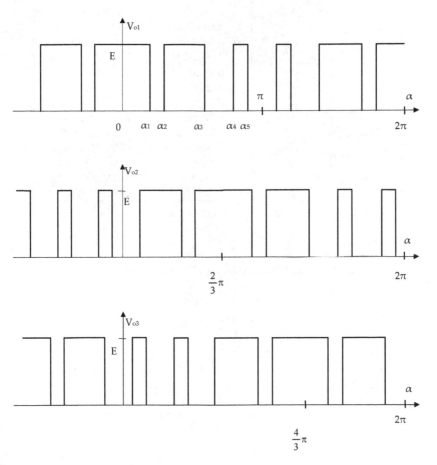

Figure 4. Inverter direct outputs representation for $N\alpha$=5.

Family	symbol	Angles (radians)	Angles (degrees)
N$_\alpha$=3	α1	0.817809468	46.8570309622392
	α2	1.009144336	57.8197113723319
	α3	1.911639657	109.528884295936
N$_\alpha$=5	α1	1.051000076	60.2178686227288
	α2	1.346257127	77.1348515165077
	α3	1.689593122	96.8065549849324
	α4	2.374938655	136.073961533976
	α5	2.47770082	141.961799882103
N$_\alpha$=7	α1	0.52422984	30.0361573268184
	α2	0.57159284	32.7498573318965
	α3	1.14918972	65.8437208158208
	α4	1.41548576	81.1013600088678
	α5	1.66041537	95.134792939653
	α6	2.16577455	124.089741091845
	α7	2.29821202	131.677849172236
N$_\alpha$=9	α1	0.43157781	24.7275870444989
	α2	0.45713212	26.1917411558679
	α3	0.70162245	40.2000051966286
	α4	0.77452452	44.3769861253959
	α5	0.96140142	55.0842437838843
	α6	1.09916539	62.9775378338511
	α7	1.21595592	69.6691422899472
	α8	1.45409688	83.3136142271409
	α9	1.64220075	94.0911720882184

Table 1. Optimal switching angles family with genetic algorithms

2.2. Dynamic LC filter behavior

Considering the inverter direct output fundamental, the LC filter transfer function is given by:

$$\overline{T} = \frac{\overline{V}_{C1}}{\overline{V}_{o1}} = \frac{1}{1 - LC\omega^2 + jRC\omega} \tag{6}$$

From (6), one can notice that for ω=0, $\overline{T}$ =1, meaning that the mean value (dc part) of the input voltage is not altered by the filter. Consequently, the inverter dc output part is entirely transferred to capacitor C_1. The same conclusion can be drawn for capacitor C_2 and C_3.

Letting $x = \omega\sqrt{LC}$ and $y = R\sqrt{\frac{C}{L}}$, the filter transfer function can rewritten as:

$$\overline{T} = \frac{\overline{V}_{C1}}{\overline{V}_{o1}} = \frac{1}{1 - x^2 + jxy} \tag{7}$$

For a given harmonic component of order k, the LC filter transfer function T_k is obtained by replacing ω with $k\omega$ as:

$$\overline{T}_k = \left(\frac{\overline{V}_{C1}}{V_{o1}} \right)_k = \frac{1}{1 - x^2 k^2 + jkxy} \tag{8}$$

Assuming that the filter L and C components are not saturated, and using the superposition principle, we obtain the inverter filtered output voltages V_{C1}, V_{C2} and V_{C3}, taken across capacitors C_1, C_2 and C_3 as given by (9.1), (9.2) and (9.3) respectively:

$$V_{C1} = E(\frac{a_0}{2} + \sum_{k=1}^{N} a_k T_k \cos(k\alpha + \varphi_k)) \tag{9.1}$$

$$V_{C2} = E(\frac{a_0}{2} + \sum_{k=1}^{N} a_k T_k \cos[k(\alpha - \frac{2}{3}\pi) + \varphi_k]) \tag{9.2}$$

$$V_{C3} = E(\frac{a_0}{2} + \sum_{k=1}^{N} a_k T_k \cos[k(\alpha - \frac{4}{3}\pi) + \varphi_k]) \tag{9.3}$$

T_k and φ_k are the kth order magnitude and phase components of the LC filter transfer function respectively.

Each k harmonic term of the inverter output voltage has a frequency of $k\omega$ and an amplitude equal to $a_k T_k$, Where a_k is the amplitude of the kth order harmonic of V_{o1}.

The transfer function magnitude and phase are given, respectively, by:

$$|\overline{T}_k| = T_k = \frac{1}{\sqrt{(1 - x^2 k^2)^2 + y^2 x^2 k^2}} \tag{10}$$

$$\varphi_k = -arctan\frac{kxy}{1 - x^2 k^2} \tag{11}$$

The maximum value T_{max} of T_k, obtained when $\frac{dT}{dx} = 0$, can be expressed by:

$$T_{max} = \frac{\frac{1}{y^2}}{\sqrt{\frac{1}{y^2} - \frac{1}{4}}} \tag{12}$$

In which case, (12) corresponds to a maximum angular frequency, this last is given by:

$$\omega_{max} = \frac{\sqrt{2}}{RC}\sqrt{1 - \frac{y^2}{2}} \tag{13}$$

The maximum angular frequency ω_{max} exists if, and only if, $y < \sqrt{2}$. The filter transfer function will exhibit a peak value then decreases towards zero. As consequence, the fundamental, as well as the harmonics, are amplified, this leads to undesirable situation, as illustrated in fig.5.

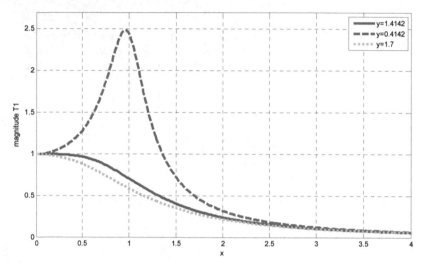

Figure 5. LC filter transfer function magnitude for the fundamental.

If (14) is satisfied, the filter transfer function will exhibit a damped behaviour.

$$y = R\sqrt{\frac{C}{L}} > \sqrt{2} \tag{14}$$

This condition matches both practical convenience and system objectives.

2.3. Harmonic rate calculation

Using (1) to (3), the non filtered inverter output voltages V_{o1}, V_{o2} and V_{o3} can be expressed as

$$V_{o1} = \frac{E}{2}(1 + \cos\omega t) + a_5 \cos 5\omega t + a_6 \cos 6\omega t + ... \tag{15.1}$$

$$V_{o2} = \frac{E}{2}[1 + \cos(\omega t - \frac{2}{3}\pi)] + a_5 \cos 5(\omega t - \frac{2}{3}\pi) + a_6 \cos 6(\omega t - \frac{2}{3}\pi) + ... \tag{15.2}$$

$$V_{o3} = \frac{E}{2}[1 + \cos(\omega t - \frac{4}{3}\pi)] + a_5 \cos 5(\omega t - \frac{4}{3}\pi) + a_6 \cos 6(\omega t - \frac{4}{3}\pi) + ... \tag{15.3}$$

Taking into consideration the filter transfer function, we get the expressions (16), (17) and (18) for V_{c1}, V_{c2} and V_{c3} respectively.

$$V_{C1} = \frac{E}{2}[1 + a_1.T_1.cos(\alpha + \varphi_1) + a_8.T_8.cos(8.\alpha + \varphi_8) + a_{10}.T_{10}.cos(10.\alpha + \varphi_{13}) + ... \tag{16}$$

$$V_{C2} = \frac{E}{2}[1 + a_1.T_1.cos([\alpha - \frac{2}{3}\pi] + \varphi_1) + a_8.T_8.cos(8[\alpha - \frac{2}{3}\pi] + \varphi_8) + a_{10}.T_{10}.cos(10[\alpha - \frac{2}{3}\pi] + \varphi_{10}) + ... \tag{17}$$

$$V_{C3} = \frac{E}{2}[1 + a_1.T_1.cos([\alpha - \frac{4}{3}\pi] + \varphi_1) + a_8.T_8.cos(8[\alpha - \frac{4}{3}\pi] + \varphi_8) + a_{10}.T_{10}.cos(10[\alpha - \frac{4}{3}\pi] + \varphi_{10}) + ... \tag{18}$$

Using (15), (16), (17) and (18), we get the three inverter filtered output voltages expressions as:

$$\begin{cases} U_{12-out} = E\frac{\sqrt{3}}{2}\left[a_1 T_1 cos(\alpha + \beta_1) + \sum_{\substack{k=8 \\ k=3n+1 \\ k=3n+2}}^{\infty} a_k T_k cos(k\alpha + \beta_k) \right] \\[3em] U_{23-out} = E\frac{\sqrt{3}}{2}\left[a_1 T_1 cos(\alpha + \beta_1 - \frac{2}{3}\pi) + \sum_{\substack{k=8 \\ k=3n+1 \\ k=3n+2}}^{\infty} a_k T_k cos(k\alpha + \beta_k - \frac{2}{3}\pi) \right] \\[3em] U_{31-out} = E\frac{\sqrt{3}}{2}\left[a_1 T_1 cos(\alpha + \beta_1 - \frac{4}{3}\pi) + \sum_{\substack{k=8 \\ k=3n+1 \\ k=3n+2}}^{\infty} a_k T_k cos(k\alpha + \beta_k - \frac{4}{3}\pi) \right] \end{cases} \tag{19}$$

with: $\beta_k = \varphi_k + \frac{\pi}{6}$.

From the precedent fig. 3, and for each lever, the equations with the currents and the voltages can be written in the following form (Bouchhida et al., 2007, 2008; Bouchhida, 2008, 2011).

- Currents equations

$$\begin{cases} \frac{dv_{C1}}{dt} = \frac{1}{C_1}(i_1 - i_{1ch}) \\[1.5em] \frac{dv_{C2}}{dt} = \frac{1}{C_2}(i_2 - i_{2ch}) \\[1.5em] \frac{dv_{C3}}{dt} = \frac{1}{C_3}(i_3 - i_{3ch}) \end{cases} \tag{20}$$

- Voltages equations:

$$
\begin{cases}
\dfrac{di_1}{dt} = (V_{o1} - R_1.i_1 - v_{C1}).\dfrac{1}{L_1} \\[2mm]
\dfrac{di_2}{dt} = (V_{o2} - R_2.i_2 - v_{C2}).\dfrac{1}{L_2} \\[2mm]
\dfrac{di_3}{dt} = (V_{o3} - R_3.i_3 - v_{C3}).\dfrac{1}{L_3}
\end{cases}
\tag{21}
$$

These equations are put in following matric form:

$$
\frac{d}{dt}\begin{bmatrix} v_C \\ I \end{bmatrix} =
\begin{bmatrix} \mathbf{0}_{3\times3} & \dfrac{1}{C}\times\mathbf{I}_{3\times3} \\[2mm] -\dfrac{1}{L}\times\mathbf{I}_{3\times3} & -\mathbf{R}\times\mathbf{I}_{3\times3} \end{bmatrix}
\begin{bmatrix} v_c \\ I \end{bmatrix} +
\begin{bmatrix} -\dfrac{1}{C}\times I_{3\times3} & 0_{3\times3} \\[2mm] 0_{3\times3} & \dfrac{1}{L}\times I_{3\times3} \end{bmatrix}
\begin{bmatrix} i_{ch} \\ V_o \end{bmatrix}
\tag{22}
$$

with : $\quad v_C = \begin{bmatrix} v_{C1} \\ v_{C2} \\ v_{C3} \end{bmatrix} ; I = \begin{bmatrix} i_1 \\ i_2 \\ i_3 \end{bmatrix} ; v_o = \begin{bmatrix} v_{o1} \\ v_{o2} \\ v_{o3} \end{bmatrix}$

3. Indirect field- oriented induction motor drive

The dynamic electrical equations of the induction machine can be expressed in the d-q synchronous reference frame as:

$$
\begin{cases}
\dfrac{di_{ds}}{dt} = -\dfrac{1}{\sigma L_s}(R_s + R_r\dfrac{M_{sr}^2}{L_r^2})i_{ds} + \omega_s i_{qs} + \dfrac{M_{sr}R_r}{\sigma L_s L_r^2}\psi_{dr} + \dfrac{M_{sr}}{\sigma L_s L_r}\psi_{qr}\omega_r + \dfrac{1}{\sigma L_s}V_{ds} \\[2mm]
\dfrac{di_{qs}}{dt} = -\omega_s i_{ds} - \dfrac{1}{\sigma L_s}(R_s + R_r\dfrac{M_{sr}^2}{L_r^2})i_{qs} - \dfrac{M_{sr}}{\sigma L_s L_r}\psi_{dr}\omega_r + \dfrac{M_{sr}R_r}{\sigma L_s L_r^2}\psi_{qr} + \dfrac{1}{\sigma L_s}V_{qs} \\[2mm]
\dfrac{d\psi_{dr}}{dt} = \dfrac{M_{sr}R_r}{L_r}i_{ds} - \dfrac{R_r}{L_r}\psi_{dr} + \omega_g\psi_{qr} \\[2mm]
\dfrac{d\psi_{qr}}{dt} = \dfrac{M_{sr}R_r}{L_r}i_{qs} - \omega_g\psi_{dr} - \dfrac{R_r}{L_r}\psi_{qr}
\end{cases}
\tag{23}
$$

$$
\frac{d\Omega_r}{dt} = -\frac{f}{j}\Omega_r - \frac{1}{j}(C_{em} - C_r)
\tag{24}
$$

$$
\Omega_r = \frac{\omega_r}{p}
\tag{25}
$$

$$
C_{em} = p\frac{M_{sr}}{L_r}(\psi_{dr}i_{qs} - \psi_{qr}i_{ds})
\tag{26}
$$

Where:

V_{ds} , V_{qs} : d-axis and q-axis stator voltages;

i_{ds} , i_{qs} : d-axis and q-axis stator currents;

ψ_{dr} ,ψ_{qr} : d-axis and q-axis rotor flux linkages;

R_s , R_r : stator and rotor resistances;

L_s , L_r : stator and rotor inductances;

M_{sr} : mutual inductance

ω_s ,ω_r : electrical stator and rotor angular speed

ω_g : slip speed $\omega g = (\omega s - \omega r)$

Ω_r : mechanical rotor angular speed

C_r , C_{em} : external load torque and motor torque

j , f : inertia constant and motor damping ratio

p : number of pole pairs

σ : leakage coefficient, $(\sigma = 1 - \dfrac{M_{sr}^2}{L_s L_r})$

Equation (23) represents the dynamic of the motor mechanical side and (26) describes the electromagnetic torque provided on the rotor. The model of a three phase squirrel cage induction motor in the synchronous reference frame, whose axis d is aligned with the rotor flux vector, ($\psi_{dr}=\psi_r$ and $\psi_{qr}=0$), can be expressed as:

$$\frac{di_{ds}}{dt} = - \gamma\, i_{ds} + \omega_s i_{qs} + \frac{K}{T_r}\psi_{dr} + \frac{1}{\sigma L_s}V_{ds} \tag{27}$$

$$\frac{di_{qs}}{dt} = - \omega_s i_{ds} - \gamma\, i_{qs} - p\Omega K\psi_{dr} + \frac{1}{\sigma L_s}V_{ds} \tag{28}$$

$$\frac{d\psi_{dr}}{dt} = \frac{M_{sr}}{T_r}i_{ds} - \frac{1}{T_r}\psi_{dr} \tag{29}$$

$$\frac{d\psi_{qr}}{dt} = \frac{M_{sr}}{T_r}i_{qs} - (\omega_s - p\Omega)\psi_{dr} \tag{30}$$

$$\frac{d\Omega}{dt} = \frac{pM_{sr}}{JL_r}(\psi_{dr}i_{qs}) - \frac{C_r}{J} - f\,\Omega \tag{31}$$

With: $T_r = \dfrac{L_r}{R_r}$, $\qquad K = \dfrac{M_{sr}}{\sigma L_s L_r}$, $\qquad \gamma = \dfrac{R_s}{\sigma L_s} + \dfrac{R_r M_{sr}^2}{\sigma L_s L_r^2}$.

The bloc diagram of the proposed indirect field-oriented induction motor drive is shown in fig.6. Speed information, obtained by encoder feedback, enables computation of the torque reference using a PI controller. The reference flux is set constant in nominal speed. For higher speeds, rotor flux must be weakened.

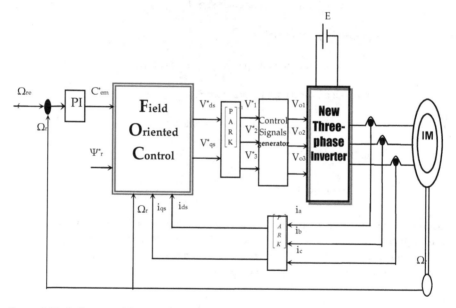

Figure 6. Block diagram of the proposed indirect field oriented induction motor drive system.

4. Simulation results

To demonstrate the performance of a new tree-phase inverter, we simulated three filtered inverter output voltages. Two frequency values are imposed on the inverter, starting with a frequency of 50 Hz, then at time t=0.06(s) the frequency is changed to 60Hz. The three filtered inverter output voltages are illustrated in (figure 7.a) which clearly shows that the three voltages are perfectly sinusoidal and follow the ideal values with a transient time of

0.012(s). The harmonic spectrum of the filtered output voltage is shown in (figure 7.b). In order to compare the performance of the new three-phase inverter with the conventional PWM inverter, the output voltage of the latter, where a modulation index of 35 was used, is shown in (figure 8.a) and its harmonic spectrum is presented in (figure 8.b).

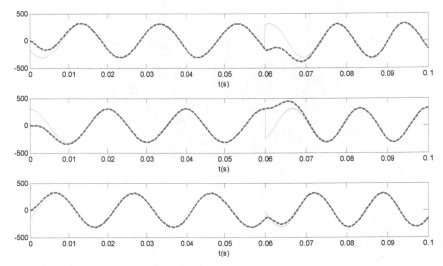

('---' filtered output voltage) ('—' ideal output voltage)

(a)

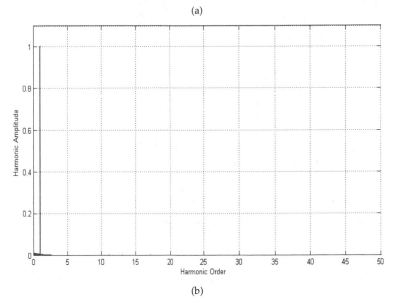

(b)

Figure 7. (a) Three filtered inverter outputs voltages.(b) Harmonic spectrum of the filtered output voltage.

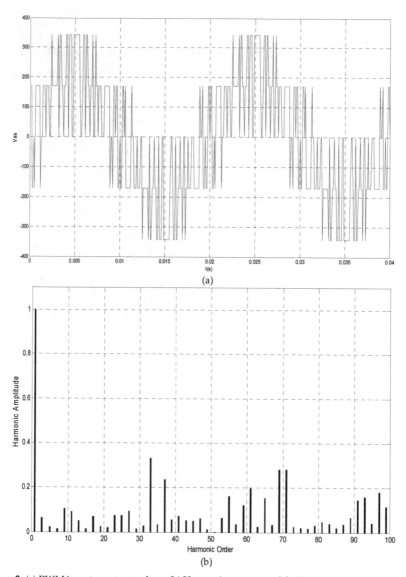

Figure 8. (a) PWM inverter output voltage (b) Harmonic spectrum of thePWM inverter output voltage.

We carried out two simulations of the field-oriented control for induction motor drives with speed regulation using the new structure of the three phase inverter in the first simulation (figure 9.a) and the conventional PWM (figure 9.b) inverter in the second simulation. The instruction speed is set to 100 (rad/sec) for both simulations. During the period between 1.3(s) and 2.3(s), a resistive torque equal to 10 (N.m) (i.e the nominal torque) is applied.

In order to illustrate the effectiveness of the proposed inverter, the torque response obtained by using the proposed and the conventional PWM inverters are shown in (figure 10.a) and (figure 10.b), respectively. The obtained results clearly show that the conventional PWM inverter generates more oscillations in the torque than the proposed structure (figure 11). Moreover, the switching frequency of the proposed inverter is dramatically reduced (see (figure 12.a)) when compared to its counterpart in the conventional PWM inverter (see figure 12.b). Therefore, the proposed inverter gives a better dynamic response than the conventional PWM inverter.

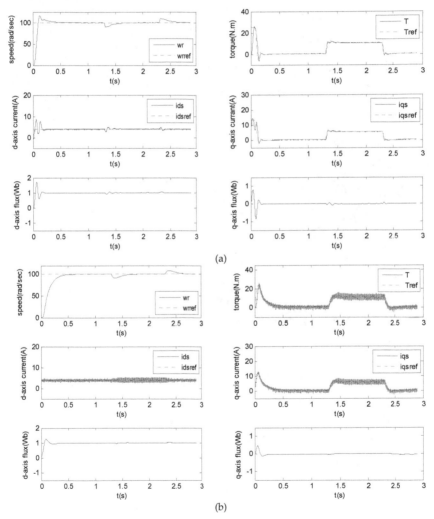

Figure 9. (a) Simulation results of the indirect field-oriented control for proposed inverters (b) Simulation results of the indirect field-oriented control for conventional PWM inverters

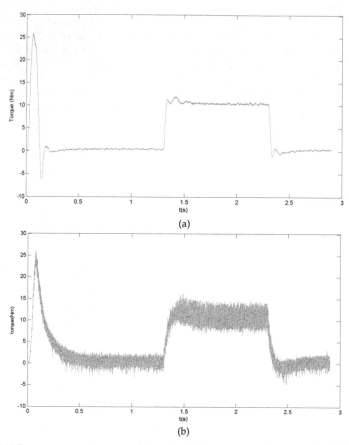

(a)

(b)

Figure 10. (a) Torque response for proposed inverters (b)Torque response for conventional PWM inverters

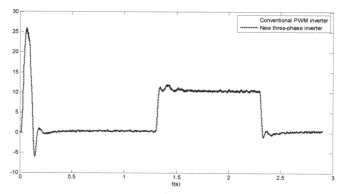

Figure 11. Torque response for proposed and conventional PWM inverters

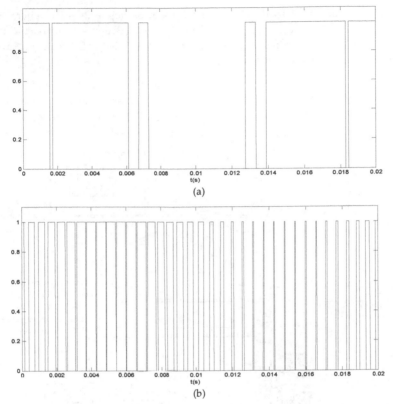

Figure 12. (a) Switching frequency for proposed inverters (b) Switching frequency for conventional PWM inverters

5. Experimental setup

The experimental setup was realized based on the DS1103 TMS320F240 dSPACE kit (dSPACE, 2006a, 2006b, 2006c, 2006d, 2006e). Figure 13 gives the global scheme of the experimental setup. This kit allows real time implementation of inverter and induction motor IM speed drive, it includes several functions such as Analog/Digital converters and digital signal filtering. In order to run the application the control algorithm must be written in C language. Then, we use the RTW and RTI packages to compile and load the algorithm on processor. To visualize and adjust the control parameters in real time we use the software control-desk which allows conducting the process by the computer.

The novel single phase inverter structure for pre-calculated switching is based on the use of IGBT (1000V/25A) with 10 kHz as switching frequency. The switching angles are predetermined off-line using Genetic Algorithms and stored in the card memory in order to speed up the programme running. The non-filtered inverter output voltages are first designed in Simulink/Matlab, then, the Real-Time Workshop is used to automatically generate optimized C code for real time application. Afterward, the interface between Simulink/Matlab and the Digital Signal Processor (DSP) (DS1103 of dSPACE) allows the control algorithm to be run on the hardware.

The master bit I/O is used to generate the required 2 gate signals, and a several Analog-to-Digital converters (ADCs) are used for the sensed line-currents, capacitors voltage, and output voltage. An optical interface board is also designed in order to isolate the entire DSP master bit I/O and ADCs. The block diagram of the experimental plant is given in figure 14

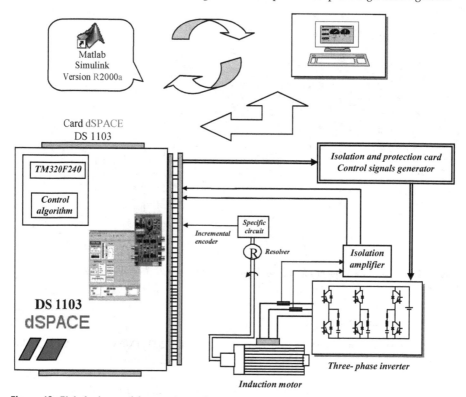

Figure 13. Global scheme of the experimental setup

6. Experimental evaluation

Figure 15 shows the experimental filtered inverter output voltage (V_{C1}-V_{C2}) for frequency value equal 50 Hz. The filtered inverter output voltage is perfectly sinusoidal. The experimental result in Figure 16 shows the torque response obtained by using the proposed PCPWM inverter: during the period between 0.65 sec and 1.95 sec, a load torque equal to 13 (N.m) is applied. The torque ripple of the induction motor is dramatically reduced.

Figure 14. Snapshot of the laboratory experimental setup

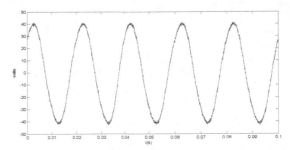

Figure 15. Experimental inverter filtered output voltage.

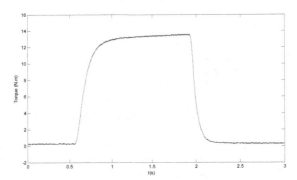

Figure 16. Experimental torque response for proposed PCPWM inverters

Moreover, as shown in Figure 17, the experimental switching frequency of the proposed PCPWM inverter is very less compared to the conventional PWM inverter one Figure 18. As a consequence, the proposed inverter provides higher dynamic response than the conventional PWM inverter in vector controlled induction motor applications.

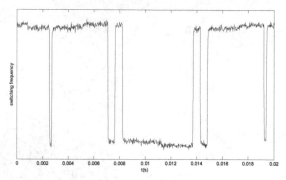

Figure 17. Experimental switching frequency for proposed PCPWM inverters

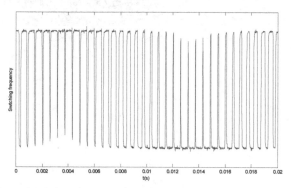

Figure 18. Experimental switching frequency for conventional PWM inverters

7. Conclusion

A three-phase inverter model was developed by combining pre-calculated switching angles and a passive filter to eliminate inverter output harmonics. The inverter model needs a nonlinear system of equations for the switching angles computation. The proposed inverter model succeeds to substantially reduce the harmonics while using polarized capacitors. The reduced number of switching angles provides more reliability and increases system components life time. Moreover, the proposed inverter design and control simplicity could be used as a cost effective solution to harmonics reduction problem. The torque ripple of the induction motor is dramatically reduced by the PCPWM inverter. The global scheme of the experimental setup has been implemented. The obtained experimental results exhibit good matching with the theoretical values. It is shown that the proposed PCPWM has better tracking performance as compared with the conventional PWM.

Author details

Ouahid Bouchhida
Université Docteur Yahia Farès de Médéa, Département Génie Electrique, Algérie

Mohamed Seghir Boucherit
Ecole Nationale Polytechnique, Département Génie Electrique, Algérie

Abederrezzek Cherifi
IUT Mantes-en-Yvelines, France

Appendix

The squirrel cage induction motor data are:

Symbol	Quantity	Value
P_n	Rated power	1.5 KW
V_n	Rated line voltage	220/380 V
C_n	Rated load torque	10 Nm
P	No. of pole pair	2
R_s	Stator resistance	5.62 Ω
R_r	Rotor resistance	4.37 Ω
M_{sr}	Mutual inductance	0.46 H
L_s	Stator leakage inductance	0.48 H
L_r	Rotor leakae inductance	0.48 H
I_n	Rated current	6.4/3.7 A
Ω_n	Motor speed	1480 tr/min
f	Viscosity coefficient	0.001136 N.m.s/rd
J	Moment of inertia	0.0049 kg.m^2
C_1, C_2, C_3	Capacitance	10 mF
L	Inductance	0.5 mH
R	Internal inductor resistance	0.5Ω

8. References

Bäck, T. (1996). *Evolutionary Algorithms in Theory and Practice*, Oxford University Press, 1996

Bouchhida, O.; Cherifi, A. & Boucherit, M.S. (2007). Novel harmonic elimination PWM Technique for reducing Torque–Ripple in Inverter-fed Induction motor. *Archives of Electrical Engineering (AEE)*, Vol. 56, No.3-4, (Mar. 2007), 197-212.

Bouchhida, O.; Benmansour, K.; Cherifi, A. & Boucherit, M.S. (2008). Low switching-frequency and novel harmonic elimination for three-phase inverter', *Proceedings of the Fifth International Multi-Conference on Systems, Signals and Devices SSD08*, IEEE, Philadelphia University Amman Jordan, July 2008.

Bouchhida, O. (2008). *Contribution à l'Optimisation de Structure des Convertisseurs pour la Commande des Machines Asynchrones: Réalisation expérimentale*. thèse de Doctorat, Ecole Nationale polytechnique ENP d'Alger, Algérie

Bouchhida, O. (2011). *Etude et Optimisation des Performances d'Onduleur Monophasé et Triphasé à Commutation Pré-Calculée*. Habilitation Universitaire, Université Saad D'hleb Blida, Algérie

Czarkowski, D.; Chudnovsky, D.; Chudnovsky, G. & Selesnick, I. (2002). Solving the optimal PWM problem for Single-phase inverters. *IEEE Transactions Circuits Syst. I*, Vol. 49, (April 2002), 465-475

Davis, L. (1991). *The Handbook of Genetic Algorithms*, Van Nostrand & Reinhold, 1991

dSPACE (2006a). *DS1103 PPC Controller Board, Feature, Realise 5.1*, dSPACE digital signal processing and control engineering, Germany

dSPACE (2006b). *How to implement user-specific functions on the DS1103 slave DSP (TMS320F240)*. dSPACE digital signal processing and control engineering GmbH, Germany

dSPACE (2006c). *DS1103 PPC Controller Board, Hardware Installation and Configuration, Realise 5.1*, dSPACE digital signal processing and control engineering, Germany

dSPACE (2006d). *DS1103 PPC Controller Board RTI Reference, Realise 5.1*, dSPACE digital signal processing and control engineering, Germany

dSPACE (2006e). *DS1103 PPC Controller Board RTLib Reference, Realise 5.1*, dSPACE digital signal processing and control engineering, Germany

García, O.; Martínez-Avial, M.D.; Cobos, A.; Uceda, J.; González, J. & Navas, A. (2003). Harmonic reducer converter. *IEEE Transactions on Industrial Electronics*, Vol. 50, No. 2, (April 2003), 322–327

Jung, J. & Nam, K. (1999). A dynamic decoupling control scheme for high speed operation of induction motors. *IEEE Transaction on Industrial Electronics*, Vol.46, No. 1, (Feb 1999), 100–110.

Lin, F.J.; Wai, R.J.; Lin, C.H.; & Liu, D.C. (2000). Decoupling stator-flux oriented induction motor drive with fuzzy neural network uncertainly observer. *IEEE Transaction on Industrial Electronics*, Vol.47, No.2, (Apr 2000), 356–367.

Meghriche, K.; Chikhi, F. & Cherifi, A. (2004). A new switching angle determination method for three leg inverter, *Proceedings of International IEEE Mechatronics and Robotics MechRob-2004*, 378–382, Aachen, Germany, September 2004

Meghriche, K.; Mansouri, O. & Cherifi, A. (2005). On the use of pre-calculated switching angles to design a new single phase static PFC inverter, *Proceedings of the 31st IEEE IECON'05*, 906–911, Raleigh North-Carolina, November 2005.

Suwankawin, S.; & Sangwongwanich, S. (2002). A speed sensorless IM drive with decoupling control and stability analysis of speed estimation. *IEEE Transaction on Industrial Electronics*, Vol. 49, No. 2, (Apr 2002), 444–455.

Villarreal-Ortiz, R.A.; Hernández-Angeles, M.; Fuerte-Esquivel, C.R. & Villanueva-Chávez, R.O. (2005). Centroid PWM technique for inverter harmonics elimination. *IEEE Transactions on Industrial Electronics*, Vol. 20, No. 2, (April 2005), 1209–1210

Wells, J.R.; Nee, B.M.; Chapman, P.L. & Krein, P.T. (2004). Optimal Harmonic elimination control, *Proceedings of the 35th Annual IEEE Power Electronics Specialists Conference*, 4214–4219, Aachen, Germany, 2004.

Chapter 9

Rotor Cage Fault Detection in Induction Motors by Motor Current Demodulation Analysis

Ivan Jaksch

Additional information is available at the end of the chapter

1. Introduction

Rotor cage faults as broken rotor bars, increased bars resistance and end-ring faults can be caused by thermal stresses, due to overload, overheating and thus mechanical stresses, magnetic stresses and dynamic stresses due to shaft torques. Environmental stresses as contamination or abrasion also contributes to the rotor cage faults. The rotor cage faults can also lead to the shaft vibration and thus bearing failures and air gap dynamic eccentricity.

Various rotor cage faults detection techniques for induction motors (IM) have been proposed during the last two decades. One of these is a widely used Motor Current Signature Analysis (MCSA) representing namely the direct spectral analysis of stator current (Thomson & Fenger, 2001; Jung et.al, 2006). MCSA can be combined with other methods as stray flux detection and a radial and axial vibration analysis. MCSA is still an open research topic, namely in the region of higher harmonics.

Strongly nonstationary working conditions as start-up current analysis require the application of methods generally called Joint Time Frequency Analysis (JTFA). These methods are Short Time Fourier Transform, Continuous Wavelet Transform (Cusido et.al, 2008; Riera-Guasp, 2008), Discrete Wavelet Transform (Kia et. al., 2009), Wigner Distribution (Blödt et al., 2008), etc. The fundamental of the wavelet analysis is the stator current decomposition into a determined number of detailed and approximation components and their pattern recognition. Wavelet analysis can be combined with other methods as a torsional vibration (Kia et. al., 2009).

The Vienna monitoring method –VMM (Kral et. al, 2008) is a rotor fault detection method based on instantaneous torque evaluation determined by voltage and current models. Other introduced methods for IM rotor fault diagnostics are multivariable monitoring (Concari et.al, 2008), artificial neural networks and neural network modeling (Su & Chong, 2007), fuzzy based approach (Zidani et. al., 2008), wavelet analysis together with hidden Markov models (Lebaroud & Clerc, 2008), pendulous oscillation of rotor magnetic field (Mirafzal & Demerdash, 2005), vibration analysis (Dorrell et al., 1977) etc. A review of diagnostic techniques has been presented in several publications (Nandi et. al., 2005; Bellini et. al., 2008; Zhang et. al., 2011).

We can see that today it is not a problem to find rotor faults. What still remains a problem is to exactly and unambiguously determine fault indicator and its fault severity with a defined measurement uncertainty under changing motor parameters - various loads and inertia. Ideally the fault indicator should be independent on IM load. The examination of the various load influence on the fault indicator changes is very important especially in the industrial applications, where the keeping the same motor loads is often difficult. It has to be known what range of parameter changes are allowed for the keeping of the fault indicator constancy. As it is not known, at repeated measurements with different fault indicators, it is not clear, if the reason is due to the varied (deteriorating) IM faults or due to IM changing parameters. What is also important is the repeatability of the measurement with the same result. The third problem is a preferably simplicity and easy implementation of the methods for practical use in industry. Introduced new demodulation methods fulfill most of these requirements.

Dynamic rotor faults of IM, namely rotor broken bars and dynamic or combined eccentricity, cause a distortion of the rotor bars current distribution, and thus they cause periodical dynamic changes related to IM rotation frequency fr in the rotor magnetic field and consequently torque oscillation and therefore stator current modulation. The complex stator current analysis and experiments based on simultaneous amplitude and phase demodulation techniques proved that the stator current at rotor faults consists both of amplitude modulation (AM) and phase modulation (PM). AM and PM are combined into the Joint Amplitude Phase Modulation (JAPM). Amplitude and phase modulating currents are in certain relations, both in their amplitudes and mutual phases, dependent on motor load, inertia and also on IM working conditions.

Motor Current Demodulation Analysis - MCDA comes out from the fundamental principle which arises at dynamic rotor faults - IM stator current modulation. The basic idea of MCDA is to extract only those currents from the whole stator current which are directly induced and caused by rotor faults, and to investigate only these fault currents in the time and frequency domain. At health IM these currents do not exist, practically are near zero.

The demodulation is a process, how to gain back the information about the time course of modulating signal from modulated signal. Outputs of demodulation methods are, therefore, the direct time courses of modulating i.e. fault currents. The spectrum of a fault current does not contain any sideband components, fault frequencies are determined directly and only one spectral peak represents the fault indicator. The great advantage of MCDA is its easy use in industrial diagnostics.

Most of the present diagnostic methods, both for stationary and non-stationary (Joint Time-Frequency Analysis) IM working conditions, use full IM current analysis methods for rotor fault detection. Just MCDA proved the complexity of the IM current at dynamic rotor faults and therefore the dependence of fault indicators based on full IM current on IM working conditions. It is mostly various IM loads which changes PM and great inertia or insufficient IM feeding which change an angle φ between AM and PM and therefore cause the dependence. These rotor fault indicators dependence cannot be removed.

2. The analysis of IM current and current signature at rotor faults

The widely used rotor fault detection technique - MCSA is presented in a large number of publications. This approach analyses whole IM stator current uses the spectral low and high slip pole frequency $2sf_i$ sidebands a_{APL}, a_{APH} around the supply frequency f_i for broken bars detection (Bellini et. al., 2001). According to this theory, magnetic rotor asymmetry causes a backward rotating field and formation of a current component a_{APL} at the low sideband frequency f_i-f_{sp}. The consequent torque and speed oscillation cause the occurrence of a new component a_{APH} at the high sideband frequency f_i+f_{sp}. First, the only low sideband spectral magnitude was taken as a fault indicator for broken bars. Later a new diagnostic index was introduced as a sum of the two spectral sideband components a_{APL}+a_{APH}.

Presented theory of the spectral low a_{APL} and high a_{APH} sidebands origination and formation and broken bars detection is quite different from MCSA theory. The theory comes from general modulation principles and exactly determinates both sidebands origination including the equations for a_{APL}, a_{APH} computation.

An exact detailed analysis of the stator current content at broken bars from the view of the complex air-gap rotor electromagnetic field analysis representing simultaneous stator current AM and PM, their mutual relation and changes at different load and inertia moment, and their formation to the MCS a_{APL} and a_{APH} magnitudes has been missing so far.

2.1. Broken bars stator current modulation and its contribution to the MCS formation

Introduced theory and rotor fault analysis and diagnostics come from the basic principle when periodical changes in rotor magnetomotive force (MMF) cause the periodical changes of IM stator current amplitude and phase, and thus the stator current AM and PM.

Broken rotor bars, as an electric fault, cause the rotor asymmetry, the distortion of the rotor current distribution, rotor current pulsation and its amplitude modulation by the slip frequency f_{slip}. Rotor bars current amplitude changes are transformed to stator current on slip pole frequency f_{sp} and appears here as a stator current AM. This modulation can be interpreted as a primary modulation.

Rotor bars current amplitude changes cause the changes in force on coils moving in a magnetic field. The force can be obtained from the vector cross product of the current vector and the flux density vector F_x= NI $\upsilon\phi/\upsilon x$, where NI is MMF, ϕ is linked magnetic flux and x is the force direction. Subsequently, the total force on a current carrying rotor coils (bars), moving in a magnetic field, changes and electromagnetic torque oscillation appears. Torsion vibration also can appear. Oscillating torque causes periodical changes in the rotating phase angle and therefore the stator current PM. This modulation can be interpreted as a subsequent or secondary modulation, so PM cannot originate without AM and JAPM always exists. The JAPM can be interpreted as a stator current modulation by a complex modulating vector which changes both its amplitude and phase. Angular speed oscillation as the derivation of phase oscillation also appears.

The frequency of periodical changes is a slip pole frequency f_{sp} (suffix sp) which is independent on the number of motor poles

$$f_{sp} = pf_{slip} = psf_{sync} = ps2f_1 / p = 2sf_1 = 2f_1 - f_r p \qquad (1)$$

For better understanding of a real IM state, which comes in IM current at dynamic rotor faults and for the explanation of a_{APL} and a_{APH} origination and formation from AM and PM, the properties of AM and PM have to be firstly well known. So the basic properties of AM and PM will be firstly presented before the explanation of the real IM state - JAMP.

2.1.1. Amplitude modulation-AM

AM is clearly visible from the time course of an IM current amplitude which is not stable, but changes according to the modulating current amplitude. It can be clearly seen as the IM current time course envelope already from about 2% deep of modulation I_{spa}/I_1 representing approximately 2 broken bars.

Spectrum of AM (suffix A) is derived from the Euler formula $cos(\omega t) = \frac{1}{2}(exp(j\omega t) + exp(-j\omega t))$ expressing the decomposition of a harmonic signal into the pair of rotating vectors (phasors) -Fig.1. Phasors' amplitudes are the half of the original harmonic signal amplitude; one rotates positive direction, the other rotates negative direction.

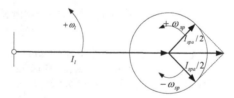

Figure 1. Vector representation of AM for broken bars

Spectra of modulated signals are always connected with two sidebands peaks around the carrier signal spectral peak. AM appears in the autospectrum of the modulated signal as three peaks: one at the carrier frequency with the magnitude equal to the amplitude of carrier signal and two sidebands spaced by modulating frequency from the carrier frequency each of them with the half amplitude of the modulating signal.

For broken bars there are two sideband components at low and high frequency $f_{L,H}$ and their low and high sidebands magnitudes a_{AL}, a_{AH} have the same size a_A which equals the half of the amplitude of the modulating current I_{spa}

$$f_{L,H} = f_1 \pm f_{sp} \qquad\qquad a_A = a_{AL} = a_{AH} = I_{spa} / 2 \qquad (2)$$

2.1.2. Phase modulation-PM

Unlike AM, PM is not observable from the stator current time course at so small modulation indexes to 0.15. PM can be visible as stator current time course compression and decompression from modulation indexes greater than 2, which are approximately 10 times higher than the usual modulation indexes at broken bars.

The spectrum of phase modulation (suffix p) is dependent on the modulation index size. If the modulation index is greater than 1, the PM spectrum consists of many side components, if it is lower than 0.4, the spectrum contains only a few side components. The spectral magnitudes computation of phase-modulated signals requires the use of the Bessel functions J_i, $i=1, 2,..,M$ (Randall, 1987).

The real modulation indexes at rotor faults, expressed by I_{spp}, are very low about to max. 0.15 at large rotor faults and in this case the stator current autospectrum contains only two significant sideband components $J_1(I_{spp})$ with the same magnitudes $a_{PL,H} = a_P$ equal to the half of the I_{spp} multiplied by phase current amplitude I_1. (Bessel functions are not needed).

$$f_{L,H} = f_1 \pm f_{sp} \qquad a_P = a_{PL} = a_{PH} = I_1 I_{spp} / 2 \qquad (3)$$

The result is that the autospectrum of PM current looks the same as the autospectrum of AM current. But the substantial difference is in the initial phases of sidebands magnitudes at $f_1 - f_{sp}$ and $f_1 + f_{sp}$ frequencies, Table 1., 2nd row, bold, and Fig.2. Spectral sideband magnitudes a_{PL}, a_{PH} increase both owing to I_{spp} and also owing to I_1, unlike at AM, see (2).

Component frequency	f_1-2f_{sp}	f_1-f_{sp}	f_1	f_1+f_{sp}	f_1+2f_{sp}
Initial phase	$\pi-2\phi$	$\pi/2-\varphi$	0	$\pi/2+\varphi$	$\pi+2\varphi$
Component amplitude	$J_2(I_{spp})$	$J_1(I_{spp})$	$J_0(I_{spp})$	$J_1(I_{spp})$	$J_2(I_{spp})$

Table 1. Frequencies and phases of the first 2 sidebands components of PM

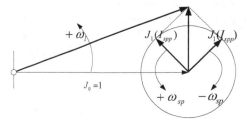

Figure 2. Vector representation of PM for broken bars

Table 1 and Fig.2 show that the initial phase of Bessel function J_1 is $\pi/2$ so the resulting vector of PM is perpendicular to the resulting vector of AM. The initial phase shift for the component at f_1+f_{sp} is the original phase shift $+\varphi$ of the modulating signal, the initial phase for the component at f_1+f_{sp} is $-\varphi$ (a highlight and bold in Table 1, middle row).

2.1.3. JAPM and MCS–the real motor state at broken bars

This type of modulation expresses exactly the real motor state at broken bars. Both AM and PM have the same frequencies and their amplitudes and phase shifts between them are in certain relations depending on IM load and inertia. Providing sinusoidal currents, the IM current of healthy motor $i_a=I_l cos(\omega_l t)$ changes at rotor broken bars to

$$i_a = \left(I_l + I_{spa} \cos \omega_{sp} t\right) \cos(\omega_l t - I_{spp} \cos(\omega_{sp} t + \varphi)) \tag{4}$$

which contains both AM and PM. Amplitude modulating current $i_{AM} = I_{spa} \cos \omega_{sp} t$, phase modulating current $i_{PM} = - I_{spp} \cos (\omega_{sp} t + \varphi)$. The term I_{spa}/I_l represents the deep of modulation of AM and I_{spp} represents the modulation index of PM.

At low inertia and normal and stable working conditions the AM and PM currents have exactly opposite phases $\varphi_0 = \pi,\ \varphi = 0$. This results from the IM torque-speed characteristic, where torque changes induced by MMF changes induce opposite changes in speed.

The necessary condition for both MCS low a_{APL} and high a_{APH} autospectral magnitudes equality is the mutual perpendicularity of vectors forming AM and PM, Fig.1, 2 and Table 1. It occurs only in the case of exactly coincident or exactly opposite phases of AM and PM.

Since the necessary condition is fulfilled, the resulting low a_{APL} and high a_{APH} autospectral sideband magnitudes of JAPM have the same size given by (5).

$$a_{APL} = a_{APH} = \sqrt{a_A^2 + a_P^2} \tag{5}$$

Unfortunately from sideband magnitudes a_{APL}, a_{APH}, which are the results of the widely used MCS, the contribution of AM and PM cannot be found out. Only demodulation techniques can find them.

Increasing IM load causes the increase of IM current I_l. Previous investigations and experiments (Jaksch & Zalud, 2010) proved also an increasing oscillation of rotor magnetic field at f_{sp} with increasing load which means the increase of I_{spp}. It means consequently the increase of a_{APL}, a_{APH} according to (3) at the same rotor fault size (dash line in Fig.3). The result is that PM is the main reason of a_{APL}, a_{APH} dependence on IM load.

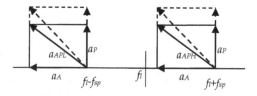

Figure 3. The formation of MCS a_{APL}, a_{APH} at broken bars from JAPM, $\varphi=0$, increasing motor load means the increase of PM –dash line.

Generally the inertia influences dynamic behavior of systems. Great inertia causes the increase of the mechanical time constant of the motor rotating system and therefore the delay $-\varphi$ (φ <0) of PM behind AM on f_{sp}. IM current spectrum symmetry and a_{APL}, a_{APH} equality disappears. The initial phase shift for the positive component at $f_i + f_{sp}$ frequency is $+\varphi$, see bold in Table I, and therefore a_{APH} decreases, the initial phases shift for the negative component at $f_i - f_{sp}$ is $-\varphi$ and therefore the a_{APL} increases -Fig.4. The angle between AM and PM modulation vectors is not $\pi/2$, (5) is not valid and the resulting a_{APL}, a_{APH}, are the vector sum of AM and PM rotating vectors.

Spectral magnitudes $a_{APL,H}$ of MCS can be computed according to the modified cosine low.

$$a_{APL,H} = \sqrt{a_A^2 + a_P^2 - 2a_A a_P \sin(\mp\varphi)} \tag{6}$$

For $\varphi=0$ this general equation for MCS computation, equation (6) changes to equation (5).

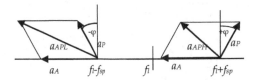

Figure 4. The formation of MCS a_{APL}, a_{APH} at broken bars from JAPM at great inertia, $\varphi= -\pi/6$

In the case of the extremely large inertia the shift between AM and PM is almost $\varphi= -\pi/2$. The result is that only the low spectral sideband component appears, see (6), with and the right sideband component is almost zero. The result is only theoretical.

The same comes in the case of overloaded and insufficiently fed IM by the voltage substantially lower than nominal AC voltage, which can be named as "abnormal working condition", when PM cannot follow AM and $\varphi<0$. In all other cases of normal IM working condition a_{APL}, a_{APH} should be the same size, in the range of possible Digital signal processing (DSP) errors.

Based on the JAMP the phenomena which appear in dynamic motor modes like start-up or breaking and their formation to a_{APL} and a_{APH} can be explained.

2.2. Dynamic eccentricity IM current modulation

The rotor cage faults can also lead to the mechanical stresses, shaft torques, bearing failures and therefore air gap dynamic eccentricity (Joksimovic, 2005; Drif & Cardoso, 2008). So the dynamic eccentricity, as a mechanical fault, closely relates to the rotor cage faults. Dynamic eccentricity is the condition of the unequal air gap between the stator and rotor caused namely by loose or bent rotor, worn bearing etc. IM current analysis is the base method, but the vibration as the cause of unbalance and the torsional vibration as the cause of the torque oscillation can be also performed as supporting method, but IM current related detection methods in most cases give very good results. Relative dynamic eccentricity is defined as the ratio of the difference between the rotor and the stator center to the difference between the stator and the rotor radius. The values of dynamic eccentricity are 0-1 or 0-100%.

Dynamic or combined eccentricity and subsequently the air gap alternation changes the rotor electromagnetic field ones per IM revolution, so the modulating frequency is the rotation frequency f_r (suffix r). IM spectrum contains two sidebands around f_l.

$$f_{L,H} = f_l \pm f_r \tag{7}$$

Spectrum of demodulated current contains peaks on direct frequencies f_r. The air gap changes do not contain stepping changes and the change usually pass subsequently during one revolution, so modulating current is often almost harmonic and contains only small higher harmonics, unlike modulating current for broken bars.

Providing sinusoidal currents, the motor current of healthy motor $i_a=I_l\ cos(\omega_l t)$ changes at rotor dynamic eccentricity to

$$i_a = \left(I_l + I_{ra} \cos \omega_r t\right)\cos(\omega_l t - I_{rp} \cos(\omega_r t)) \tag{8}$$

which contains both AM and PM.

2.3. The influence of the time varying load

Until now the modulations caused by the internal IM rotor faults were solved providing IM different, but constant load when MCDA spectrum contains two significant peaks on f_{sp} and f_r , see Fig.6. In the case of the external periodical harmonic time varying load, which varies with the frequency $f_{load} < f_l$, both additional AM and PM of IM current arise on the f_{load} frequency. It can come e.g. in the case when IM drives machines with various machine cycles e.g. textile machines, machine tools etc. (In the case of the speed reducing devices as gearbox transmissions, the gear-ratio has to be counted for f_{load} determination).

In the IM full current spectrum - MCS two additional spectral sidebands peaks appear on frequencies $f_l \pm f_{load}$. In MCDA spectrum the time varying load appears as a one spectral peak on f_{load} with the magnitude proportional to a load torque difference which can be expressed as an I_{load}. So together 3 significant spectral peaks on f_{sp}, f_r and f_{load} appear in MCDA spectrum of IM with rotor faults. The AM can be also observed in the time course of IM current as the stator current envelope or in the time course of amplitude demodulated current.

If f_{load} equals exactly f_{sp} or f_r the rotor faults diagnostics is not correct because the resulting spectral magnitudes are the vector sum of the corresponding modulating amplitudes. The situation when $f_{load} = f_r$ can come when IM directly drives a machine with uneven load during one revolution e.g. a cam mounted on the main shaft. But practically it is a minimal probability that f_{load} equals f_{sp} or f_r. The minimal difference between f_{load} and f_{sp} or f_r is at Hanning window approximately 4 discrete step Δf. Usually used acquisition time is $T= 4s$, $\Delta f=1/T= 0.25$ Hz, so a minimal difference between external f_{load} and rotor faults frequencies f_{sp},

f_r is roughly 1Hz for the rotor fault diagnostic not influenced by the time varying load. The f_{sp} computation from f_r (1), enables the differentiation and subsequently the omission of an additional disturbing spectral peak on f_{load}.

Small external time varying load can be also caused by the fault of IM driven devices as a gearbox. In MCDA spectrum another peak appears and by the allocation of this peak frequency to the gearbox relevant mechanism frequency, the device fault can be identified.

3. Demodulation methods for rotor faults

Generally, the demodulation methods extract the original AM and PM signals using special computation methods. Now the demodulated signals are original modulating signals.

Demodulation methods can be used without precarious presumption whether the signal is modulated or not and in the case of no modulation, the demodulation results are zeros (PM) or constants (AM) and by removing DC also zeros.

For the determination of the modulated signal instantaneous amplitude and phase, a complex analytical signal has to be defined and created. An analytical complex signal created by mathematical formula is the base for the demodulation analysis. The most used methods are Hilbert transform, Hilbert-Huang transform or quadrate mixing. For 3-phase motors rotor faults a new method based on the space transform was developed.

The rotor fault amplitude demodulation extracts the original AM current. The AM current appears in stator current as an envelope of this current -Fig.5. Therefore the amplitude demodulation is known as an envelope analysis (Jaksch, 2003). It is the base for dynamic rotor fault diagnostics.

At broken bars, phase demodulation extracts PM current I_{spp} [A] which, as an argument of harmonic function (4), really represents the phase angle ripple or phase swinging [rad]. The phase demodulation gives the time course of the instantaneous swinging angle or instantaneous angular speed and represents a huge tool for the research of the rotor magnetic field oscillation, sensor-less angular speed, speed variation or other irregularities.

The demodulation analysis should be used for band pass filtered signals with the center in a carrier frequency and span corresponding to the maximal modulating frequency. Spectrum of demodulated current outside this bandwidth is shifted by a carrier frequency towards to the low frequencies. For IM rotor faults the carrier frequency is usually a supply frequency f_i and the maximal modulating frequency is f_r, so the basic bandwidth $0-2f_i$ is suitable. In the case when analog bandwidth $0-2f_i$ cannot be kept it is possible to use higher bandwidths, but Shannon sampling theorem has to be strictly kept and demodulated spectrum must be evaluated only in the range of $0-f_i$ because for higher frequencies is not valid.

Higher order harmonics of supply current and also modulating broken bar current should appear as sideband components at frequencies $f_{k,l} = kf_i \pm 2lsf_i$, k=3,5,7, l= 1,2,3 where k represents the index for stator current harmonics and l represents the index for broken bar sideband current harmonics. Because of the interaction of time harmonics with a space harmonics, a saturation related permeance harmonics together with phase shifts of AM and PM means that some sidebands harmonics are suppressed and only certain ones can appear, so the above introduced formula is not generally valid.

Demodulation in the region of higher k-harmonics of the supply frequency requires the shift of the supply carrier frequency kf_1 to zero before the demodulation. It means the spectral frequency resolution Δf increasing which is often called Band Selectable Fourier Analysis (BSFA) or Zoom. Dynamic signal analyzers are equipped with this function (zoom mode) and a maximal $\Delta f = 1mHz$. However, this analysis has a little practical sense, because of above mentioned problems with higher order harmonics. In addition the modulating currents there usually have smaller amplitudes.

3.1. The demodulation using Hilbert transformation

The Hilbert transform (Bendat, 1989) is a well-known tool which enables to create an artificial complex signal $H(t)= x(t)+jy(t)$, called analytical signal, from a real input signal $x(t)$. The real part $x(t)$ of the analytical signal $H(t)$ is the original signal – stator current, the imaginary part $jy(t)$ represents the Hilbert transform of a real part $x(t)$. The absolute value $magH(t)$ representing amplitude demodulation and the phase $\beta(t)$ representing phase demodulation can be computed according to (9), (10).

$$magH(t) = \sqrt{x^2(t)+y^2}(t) \qquad (9)$$

$$\beta(t) = \arctan(y(t)/x(t)) \quad in < -\pi, \pi > \qquad (10)$$

The amplitude demodulation computation is quite easy and can be used for continual monitoring and diagnostics in real time.

Phase determination $\beta(t)$ [rad] from a complex number position holds only in the range $<-\pi, \pi>$ and if the phase overlaps these limits, it is necessary to unwrap it and moreover to remove the phase trend component which increases by 2π every revolution with the increase of the common phase carrier signal (Randall, 1987). The computation of phase demodulation is a little difficult comparing to amplitude demodulation, but generally there is usually no problem with the phase demodulation computation.

3.2. The demodulation using space transformation

The space (Park) transform, based on the physical motor model, is used primarily for the motor vector control. A three phase i_a, i_b, i_c, $i_a(t)=I_1\cos(\omega t)$ system is expressed in one space current vector $i= K_s (i_a + a\ i_b + a^2 i_c)$, $a= e^{j2\pi/3}$ projected to the complex d-q plane (11).

From the 3 possible choices of K_s: $K_s = 1$ - amplitude invariance, $K_s = sqrt(2/3)$- power invariance, $K_s=2/3$ is used. In this case the additional recomputation coefficient between phase currents and transformed currents does not have to be used.

$$i_d(t) = (2/3)i_a(t) - (1/3)i_b(t) - (1/3)i_c(t)$$
$$i_q(t) = (\sqrt{3}/3)i_b(t) - (\sqrt{3}/3)i_c(t) \tag{11}$$

Space transform was firstly also used for the demodulation (Jaksch, 2003). From the viewpoint of the means necessary for the demodulation process, space vector $P(t)=i_d(t)+ji_q(t)$ represents a complex analytical signal computed from three three currents i_a, i_b, i_c similarly like the Hilbert transform creates the artificial complex signal $H(t)$ from one phase current.

The absolute value $magP(t)=sqrt(i_d{}^2(t)+i_q{}^2(t))$ forms the amplitude demodulation, $\beta(t) = arctan\ (i_q(t)/\ i_d\ (t))$ forms the phase demodulation.

3.3. The comparison of both demodulation methods

The space transform requires 3 currents measurement, but only simple computation (11) and no other transformation for the complex analytical signal determination. On the contrary Hilbert transform needs only one current measurement, but $jy(t)$ computation.

The space transform creates the analytical signal from 3 currents. In order to obtain the same sizes of fault indicators as from Hilbert transform, (11), using $K_s=2/3$ must be kept.

Small differences between Hilbert and space transforms can occur in the following cases:

- The violation of the exact phase shift $a=e^{j2\pi/3}$ between IM phase currents (space transform supposes exact shift $2\pi/3$).
- Power feeding voltage unbalance or great stator fault, which can cause greater IM currents unbalance.

The maximum error should not be greater than in the range of several percent. The experiment showed - see Table II., Table III, I_{spaH}, I_{spaP} that the differences are up to 5 %.

Amplitude demodulation can be implemented also by the other techniques resulting also from the three phase IM feeding system as an apparent power magnitude or a squared stator current space vector magnitude. However these methods are more complicated than the space transformation method and in addition the results of these methods are in units and dimensions which are not comparable with the Hilbert transform results.

4. Simulation results

Various simulations have been performed. The main aim of the simulation was the verification of (5), (6) for the IM current MCS - a_{APL}, a_{APH} computation, namely the influence of angle φ on the sizes of a_{APL}, a_{APH}. The verification of the equality of MCDA fault indicators (lower window in Fig. 5) with input data values I_{spa}, I_{ra} also has been performed.

As it was previously derived in the section 2, the IM current of healthy motor $i_a=I_1\ cos(\omega_1t)$ changes at dynamic rotor faults - rotor broken bars and dynamic eccentricity to

$$i_a = \left(I_1 + I_{spa}\cos\omega_{sp}t + I_{ra}\cos\omega_r t\right)\cos(\omega_1 t - I_{spp}\cos(\omega_{sp}t + \varphi) - I_{rp}\cos(\omega_r t)) \tag{12}$$

Input data for the simulation result from (12). Simulation values of I_l, I_{spa}, I_{spp}, I_{ra}, I_{rp}, φ start from the measurement, but various values can also be used. Other data processing is the same as in experiments. Hilbert transform, (9), (10) was used for the IM current amplitude and phase demodulation. The values for a_{APL}, a_{APH} (2nd window in Fig.5) were compared to the values computed from (3),(6). Full identity with the theory was found.

The simulation was also used for the case where angle φ is positive and $a_{APL} < a_{APH}$. But this IM state is not stable and can come only in IM dynamic regime.

Simulation results are depicted in Fig.5. Note that the time course of amplitude demodulated current follows the envelope of IM current – compare the window 3 to the window 1. Time course of amplitude and phase demodulation (windows 3, 4) shows small ripple at the beginning - t=0, given by nonsequenced modulating current in time window.

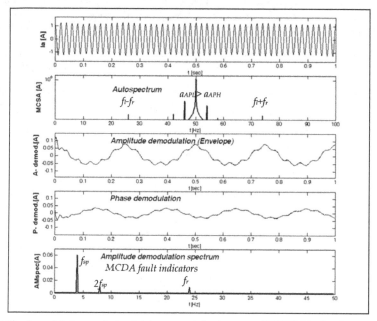

Figure 5. The demodulation analysis of stator current, 4-poles motor 0.75 kW, 2 broken bars, great inertia - $\varphi < 0$, low dynamic eccentricity. Simulation results.

5. Experimental results

The main aim of the experiments was to verify the introduced theory of JAPM, to verify the both fault indicators I_{spa}, I_{ra} changes at different IM load and to verify the influence of the time varying load.

Two sorts of IM were used for the experiments.

The first IM was SIEMENS type 1LA7083-2AA10, 1.1 kW, two-pole, rated revolutions 2850 min^{-1}, I_{nom}=2.4A, n_{rb} =23, air gap dimension=0.25 mm, health motor, 1 interrupted rotor bar and 2 contiguous interrupted rotor bars, both 3 rotors were balanced with factory set-up dynamic eccentricity (setting the exact value of dynamic eccentricity is at so small air gap very difficult). The 2nd IM was SIEMENS type 1LA 7083-4AA10, 0.75 kW, four-pole motor, I_{nom} = 1.8A, n_{rb} =26, balanced rotor and rotor with 2 contiguous interrupted rotor bars.

Various motors and fault rotors used in experiments were manufactured directly at Siemens Electromotor. Motors were tested at 25%, 50%, 75% and 85% of the full load according to the motor load record from Siemens. The changes in the broken bar fault indicator I_{spa} was also tested in the low load range from no load to 20% of full load.

The experiments were based on Bruel&Kjaer PULSE 20 bits dynamic signal analyzer (DSA) based on the frequency filtration and decimation principle. All channels are sampled simultaneously. FFT analyzer was set on the base band mode, frequency span 100Hz, 400 frequency lines, Δf =0.25Hz, Hanning window, continuous RMS exponential averaging with 75% overlapping. For the experiment of I_{spa} changes at very low load the measuring time T= 32s, Δf =0.03125Hz was used.

To find out the possible differences in both introduced demodulation methods, both Hilbert and space transform were simultaneously evaluated in the real time.

The experiments results were verified by 16 channels PC measurement system based on two 8 channels, 24-bit DSA NI 4472B from National Instruments setting in the lowest possible frequency range 1kHz.

To obtain the maximal measurement accuracy the possible errors in Digital Signal Processing (DSP) should be avoided. Sampling theorem with full agreement between the sampling frequency and surveyed analog frequency band should be strictly kept. At the violation of sampling theorem, signal frequencies higher than Nyquist frequency f_N are tilted - masked to the basic frequency region from 0-f_N and they can create there aliasing frequencies or interfere with regular frequencies, changing their amplitudes. Masking can come through many higher bands of the sampling frequency. Unlike dynamic signal analyzers, simple PC cards and scopes are usually not equipped with anti aliasing filters.

The measurement acquisition time T should be optimally set. Spectral frequency resolution Δf is a reciprocal value of the acquisition window T, Δf =1/T.

A great DSP error, both in frequency and magnitude, occurs if the analog frequency of the examined signal is exactly in the centre of Δf. In the case of a rectangular window, the spectral magnitude decline is $sinc(\pi/2)=2/\pi=0.636=-3.92$ dB, representing 36% fault in magnitude! In the case of Hanning window the decline is $(3/\pi -1/(3\pi))=0.848=-1.43$ dB. If the low analog sideband frequency f_i-f_{sp} is nearer to the discrete spectral frequency than the high sideband frequency f_i+f_{sp}, the a_{APL} can be higher than a_{APH} and vice versa. The optimal acquisition time should be longer than 1sec. e.g. 4sec. with Δf = 0.25Hz. In the case of very low load, the minimal acquisition time 8 sec with Δf = 0.125Hz can be used for the accurate f_{sp} detection and for the decrease of DSP errors probability.

Spectrum averaging, which lowers stator current non-stationary errors, should be always used for the error minimization. FFT computation time is substantially shorter than the acquisition time, so the start of a new acquisition and a new averaging can start earlier than the end of the previous acquisition time. This process is called overlapping. It is expressed in percent of the acquisition time in the range of 0% - no overlapping- to max, when the new acquisition starts immediately at the end of previous FFT. The overlapping implementation (programming) is easy. Overlapping more than 50% is recommended.

Spectrum of demodulated current does not contain any sidebands components, it is transparent and easy readable and only one simple spectral peak is the fault indicator.

The fault indicator I_{spa} [A] for broken bars is the amplitude of the amplitude modulating current on fault frequency f_{sp} so the spectral magnitude of amplitude demodulated IM current on f_{sp}, see Fig.6, 4rd window from the top. The fault indicator I_{ra} [A] for dynamic eccentricity is the amplitude of the amplitude modulating current on fault frequency f_r so the spectral magnitude of amplitude demodulated IM current on f_r, Fig.6, 4rd window.

Fault indicators clearly show the rotor faults but do not show the real fault severity. I_{spa} and I_{ra} amplitudes considerably differ with the IM power.

Fault severity dimensionless coefficients k_{sp}, k_r [%] are fault indicators normalized by a constant value - motor rated current I_{nom}

$$k_{sp} = I_{spa} / I_{nom} \qquad\qquad (13)$$

$$k_r = I_{ra} / I_{nom} \qquad\qquad (14)$$

where fault indicators I_{spa}, I_{ra} are expressed in RMS.

In order to keep the independence of fault severity coefficients of the different load, the normalizing value must be a constant. Therefore the coefficients k_{sp}, k_r as the basic evaluating tool for the assessment of fault severity and for the state of rotor bars was suggested.

Five various experiments, presented in paragraphs 5.1 to 5.5, covering different IM states, time varying load and IM energized from inverters were performed.

5.1. The results of MCDA fault indicators for broken bars at different IM load

The complex stator current analysis using simultaneous amplitude and phase demodulation of the two-pole IM is depicted in Fig. 6 – individual windows from above: stator current autospectrum i.e. MCS, AM fault current extracted from the stator current by the amplitude demodulation (9), PM fault current extracted from the stator current by the phase demodulation (10). Corresponding spectra are depicted in the 4th and 5th windows. The time courses of AM and also PM clearly show fault currents with 2 dominating frequencies f_{sp} and f_r which are better seen in the corresponding spectra. It is clearly seen that the phases of AM and PM at f_{sp} are exactly opposite, compare 2nd window to 3rd window from the top.

MCDA fault indicators are depicted in 4th window from the top.

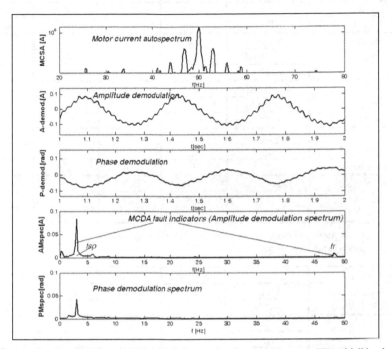

Figure 6. Complex demodulation analysis of IM current, 2-poles 1.1.kW motor, 75% of full load

Demodulation methods suppress large I_l and therefore the relatively accurate linear scale for spectrum can be used, but for the observation of higher harmonics of f_{sp} a logarithmic scale also can be used.

Equation (5) which pays only under steady conditions and low inertia, was verified. The full agreement with the theory of a_{APL} and a_{APH} equality was found in the range of possible a_{APL}, a_{APH} variations, which can be in the range of $\pm$ 2.5dB.

There are two reasons for the a_{APL} and a_{APH} variation:

1. The non-stationarity of IM current. The rotor analog current is not a fully stationary signal and therefore the phase shift φ between AM and PM may not be exactly zero and due to small dynamic changes it can oscillate around zero and therefore causes changes in a_{APL} and a_{APH} according to (6). Linear or exponential spectrum averaging lowers this error.
2. DSP errors. DSP errors come owing to the finite Δf which was described in detail in the previous part of this section 5.

The shortened experimental results are briefly summarized in Table 2 (suffix$_H$ for the demodulation using Hilbert transform, suffix$_P$ for the demodulation using space transform) for 2-pole motors and Table 3 for 4-pole motors.

Motor state		25% load	50% load	75% load	85% load
2-poles	f_r [Hz]	48.95	48.6	48.1	47.6
1.1.kW	f_{sp} [Hz]	1.99	2.8	3.8	4.62
Health	I_{spaP} [mA]	9.4	9.5	9.3	8.5
motor	I_{spaH} [mA]	9.5	9.6	9.6	8.2
	k_{sp} [%]	0.28	0.29	0.28	0.25
1	I_{spaP} [mA]	35	40	39	39
interrupted	I_{spaH} [mA]	36	41	40	39
rotor bar	k_{sp}[%]	1.11	1.28	1.26	1.25
2 contin.	I_{spaP} [mA]	69	78	79	77
interrupted	I_{spaH} [mA]	67	75	77	76
rotor bars	k_{sp}[%]	1.93	1.98	1.95	2.01
Dynamic ecc.	I_{raP}[mA]	3.9	3.9	3.9	3.5
Balanced	I_{raH}[mA]	3.8	3.9	4.1	3.3
rotor	k_r[%]	0.15	0.15	0.16	0.13

Table 2. Fault Indicators and fault severity coefficients for broken bars and dynamic eccentricity, Hilbert and Space Transforms, 2- poles IM

Motor state		25% load	50% load	75% load	85% load
4- poles	f_r [Hz]	24.56	24.41	24.21	24.12
0.75 kW	f_{sp} [Hz]	1.73	2.3	3.1	3.45
Health	I_{spaP} [mA]	3.3	3.4	3.4	3.3
motor	I_{spaH} [mA]	3.4	3.3	3.4	3.2
	k_{sp} [%]	0.18	0.18	0.18	0.17
2 contin.	I_{spaP} [mA]	37	41	43	42
interrupted	I_{spaH} [mA]	38	42	43	41
Rotor bars	k_{sp}[%]	1.46	1.68	1.70	1.68
Balanced	I_{raP}[mA]	5.1	5.4	5.3	4.9
rotor	I_{raH}[mA]	5.2	5.5	5.4	4.9
	k_r[%]	0.28	0.28	0.27	0.26

Table 3. Fault Indicators and fault severity coefficients for broken bars and dynamic eccentricity, Hilbert and Space Transforms, 4- poles IM

5.1.1. The experimental results discussion

The experiments proved the correctness of JAPM theory and the correctness of the used demodulation techniques.

The experiments show that broken bar AM is almost insensitive to the motor load. The I_{spa} changes are in the range of 11% within the interval between 25 – 85% and in the range of 7 % in the interval 50 – 85%. The same holds for the fault severity coefficient k_{sp} because the denominator in (13) is a constant value. The values of fault indicator for 2 continuous rotor bars are almost twice greater then indicators for 1 broken bar.

At no load and at low load below 20 % of full load I_{spa} decreases- see Table 5. Therefore for the industrial diagnostics the recommended load range is from 20% of load to the full load. For more accurate diagnostics, the range from 25% of load to the full load is recommended.

Fault severity coefficient k_{sp} for 2 broken bars at 4-poles motor are little smaller in comparison with 2-poles motor owing to a greater number of rotor bars n_{rb} - 26 on 23.

Healthy motor shows some residual modulation, due to irregularities in rotor bars layout, but k_{sp} were not at all measurements greater than 0.3 %.

The acceptable limits for k_{sp}, should be experimentally stated for various motor types because they can differ. Namely the sizes of k_{sp} for large IM can have different acceptable limits.

The values of fault severity coefficient k_r for dynamic eccentricity slightly decrease with increasing load. Over the 85% of full load the decrease is slightly greater. The values of I_{ra} at a factory balanced rotor are still sufficient for a sensorless rotor electromagnetic field speed and speed irregularities measurement (Jaksch & Zalud, 2010). PM for rotor eccentricity slightly decreases with increasing load unlike PM for broken bars.

Two introduced demodulation methods for dynamic rotor fault detection - Hilbert and space transforms - give the same results and both can be used for rotor fault diagnostic. Both measurement systems - Bruel&Kjaer PULSE and NI 4472B give the same fault indicators.

5.2. The comparison of AM and PM at different loads

The second experiment examined the spectral magnitudes of amplitude and phase demodulated IM current on f_{sp} at different loads, together with the comparison of the MCSA low sideband a_{APL} on f_i-f_{sp}. The results are depicted in Fig.7 -9 and summarized in Table 4.

2-poles IM, 1.1 kW 2 broken bars	25% load	50% load	75% load	85% load
AM - MCDA fault indicator I_{spa} [mA]	70	77	78	78
PM [mrad]	24	33	44	49
Low sideband a_{APL} [mA]	36	44	46	48

Table 4. AM and PM, 2 broken bars, at different IM load in comparison to low sideband a_{APL}

The experiment proved the theory that PM substantially increases with increasing load (2nd row in Table 4 and peaks in circles on f_{sp} in lower window in Fig.7 to Fig.9).

PM increases 2.2 times within the interval between 25 - 85%. The increase of PM spectral magnitude a_P is caused both by I_l increase at increasing load and also by I_{spp} increase, (3). It is the real cause of MCS fault indicators a_{APL} and a_{APH} increasing with increasing load, Fig.3 and (5).

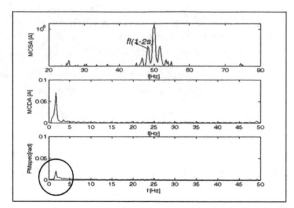

Figure 7. IM spectrum, spectrum of amplitude and phase demodulated current, 25% of full load

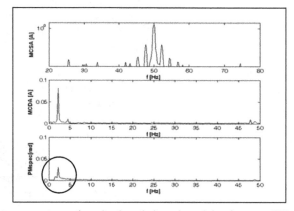

Figure 8. IM spectrum, spectrum of amplitude and phase demodulated current, 50% of full load

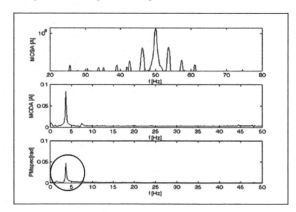

Figure 9. IM spectrum, spectrum of amplitude and phase demodulated current, 85% of full load

5.3. The analysis of indicator I_{spa} at very low load from no load to 20 % of full load

The 3rd experiment analyses the changes of I_{spa} in the range of no load to 20 % of full load. Table 5 shows the MCDA broken bars fault indicator I_{spa} decline under 20% of full load. For very low load at s=0.44%, f_{sp}=0.44 Hz the I_{spa} for 2 broken bars decreases to I_{spa}=31mA, which is approximately the half of its nominal value and it corresponds to I_{spa} for 1 broken bar (Table 2), so great confusion in broken bar diagnostics may come.

s[%]	0.21	0.37	0.44	0.69	0.94	1.16	1.37
f_{sp} [Hz]	0.21	0.37	0.44	0.69	0.94	1.16	1.37
I_{spaH} [mA]	19	25	31	37	47	56	69

Table 5. I_{spa} changes from no load to 20% of full load, 2 broken bars, 1.1.kW IM

The decrease of I_{spp} representing PM, and therefore the decrease of MCSA fault indicators a_{APL}, a_{APH} under 20% are substantially faster than the decrease of MCDA fault indicator I_{spa} (Fig.7–Fig.9, PM in circles).

5.4. IM rotor fault diagnostics at time varying load

The aim of 4th experiment was to verify presented theory of the time varying load and its influence on MCDA fault indicator I_{spa}. The time varying load frequency f_{load} = 4 Hz was chosen very near to the broken bars fault indicator frequency f_{sp} =1.9 Hz.

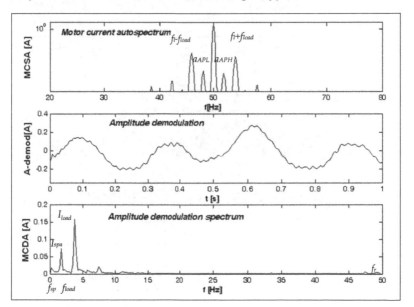

Figure 10. Windows from above: MCS, time course of amplitude demodulated current, and its MCDA spectrum, 25% of full load.

Experimental results are depicted in Fig.10. Additional two new spectral sidebands appear on frequencies $f_l \pm f_{load}$ in MCSA spectrum-upper window and one new spectral peak appears on f_{load} in MCDA spectrum –lower window.

The time course of amplitude demodulated current, so the time course of amplitude modulating (i.e. fault) current is depicted in the middle window. The sum of three harmonic with the fundamental amplitudes I_{spa}, I_{load} and I_r and corresponding frequencies f_{sp}, f_{load} and small ripple from f_r is clearly visible.

The broken bar fault indicator I_{spa} did not change its amplitude (compare with Fig.7). The important conclusion is that the size of MCDA fault indicator I_{spa} is not influenced by the time varying load.

5.5. IM energized from inverters

The last experiment examined two-pole IM with 2 broken bars energized from inverters. Two PWM open-loop inverters were used. The measurements were made at low inertia and at a stable motor load.

In the first measurement, an inverter was properly rated. In this case, the results correspond to the results for IMs energized from line. The MCS $a_{APL} = a_{APH}$ and time course of AM and PM was completely opposite.

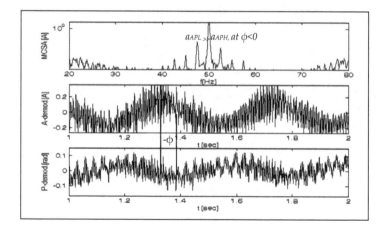

Figure 11. Inverter fed IM, low DC-link and overloaded, low inertia, 75% of full load, MCS, time course of amplitude and phase demodulated current, PM delays -φ behind AM.

In the second measurement another inverter was not properly rated and its DC-link voltage was only 400V. IM was fed by lower voltage apps. 185V and therefore was overloaded. In this case the MCS $a_{APL} > a_{APH}$ appeared (the same phenomenon which can appear at great inertia). The demodulation detected PM delay $-\varphi$ behind AM -see Fig.11 and validated the explanation of $a_{APL} > a_{APH}$ in Fig.4., and also the correctness of (6).

This phenomenon of PM delay behind AM at low inertia appears when an AC source is not properly dimensioned, its feeding voltage is lower than nominal voltage and therefore it is not able to fully excite IM.

It can be concluded that continual increasing of a_{APL} above a_{APH} (out of the range of possible a_{APL} and a_{APH} variations± 2.5dB) generally means, that IM working conditions are out of the normal.

6. Demodulation analysis versus direct analysis

The presented MCDA and following AM and PM synthesis to the full IM current fault indicators a_{APL}, a_{APH} (6) enables the comparison of the demodulation analysis to the direct full IM current analysis.

The IM stator current demodulation analysis enables to find out the complex changes in the rotor electromagnetic field and MMF. Based on this analysis the new method MCDA was introduced. MCDA comes from the basic fundamental which occurs at dynamic rotor faults – JAPM. MCDA senses the whole stator current, but before the further evaluation as spectral analysis, it extracts time courses of fault currents directly induced by dynamic rotor faults. Anything more accurate than the extraction and direct processing of fault currents cannot exist.

The phase demodulation extracts PM current and can be used for the research of rotor magnetic field changes and oscillation and for the sensorless speed measurement. PM is not very suitable for rotor fault diagnostics, because the fault indicators are dependent on motor load.

The amplitude demodulation extracts AM fault current. AM is from 20% of full load almost independent of motor load and it is the base for rotor fault diagnostics not dependent on different load and inertia. Broken bar and dynamic eccentricity fault indicators are simple spectral peaks at direct fault frequencies f_{sp}, f_r (no sidebands). The amplitude demodulation can be easily implemented continually in a real time.

MCDA is a clear, very simple and reliable method and it is very useful for industrial application both for diagnostics and also for IM continual monitoring.

Methods of the direct analysis of IM current sense and subsequently process the full IM current. The great disadvantage of the direct IM current analysis is that its fault indicators a_{APL}, a_{APH} are dependent on IM load and inertia moment. The second disadvantage is that fault frequencies cannot be determined directly, but as a difference from f_i, which can change. The third disadvantage is the lower resolution both in frequency and amplitude.

Great differences between the magnitudes of the main spectral peak on f_i and the sidebands magnitudes a_{APL}, a_{APH} requires the use of logarithmic or dB scale.

The presented demodulation analysis of IM current proved that IM current at dynamic rotor faults is not so simple and inwardly contains JAPM. Simply to say stator current consists of 3 parts - stator current of health motor, amplitude modulating current and phase modulating current (12). Just JAPM is the main reason for of the full current based fault indicators dependences on motor load and inertia.

The demodulation analysis exactly established the reasons for a_{APL} and a_{APH} formation and developed equations for their computation (5), (6). Consequently the MCDA has allowed a complete explanation of both MCS load and inertia dependences: MCS fault indicators a_{APL}, a_{APH} actually consist of 3 variables - AM, PM and φ, (6). PM increases with the increasing load (Fig.7-9, Table 4) and therefore causes the increase of both a_{APL} and a_{APH}. Great inertia or poorly fed IM causes that the angle φ changes from zero values to negative values and therefore $a_{APL} > a_{APH}$ (6). The summation or averaging of a_{APL} and a_{APH} is inaccurate, because the equation (6) is not a linear function.

The processing and direct analysis of the whole stator current does not enable the distinction of the individual AM, PM and φ contribution to the a_{APL} and a_{APH} and the result is the continual dependence of MCSA a_{APL} and a_{APH} fault indicators on IM load and inertia. No improvements and sophistication of the measurement and evaluation methods can reduce this dependence. Logarithmic or dB scale has to be used for a_{APL} and a_{APH} displaying. It together means the low resolution both in amplitudes and frequencies.

7. Conclusion

Dynamic rotor faults cause dynamic changes of rotor electromagnetic field and MMF and therefore the motor current AM and PM together creating JAPM.

Based on this theory, the new diagnostic method MCDA was developed. Basic properties of AM and PM at rotor faults were presented. The analyses of IM current and current signature at rotor faults together with mathematical equations for low and high autospectral magnitudes computation were derived. These equations were verified by simulation and experimentally and the possible low and high autospectral magnitudes variation due to IM current nonstationarity and DSP errors were discussed.

Two demodulation methods based on Hilbert transform and space transform are presented. Space transform gives the same results as Hilbert transform.

The phase demodulation extracts the fault PM current. PM can be used for the research of rotor magnetic field phase changes and oscillation and for the sensorless rotor magnetic field speed measurement. PM fault indicators are dependent on the motor load.

The amplitude demodulation extracts the fault AM current. The simple spectral peaks at fault frequencies f_{sp} and f_r are the dynamic rotor faults indicators almost independent on IM load and inertia. It was also proved that MCDA fault indicators are not influenced by the time varying load.

The presented diagnostic method is very clear and can be easily used for a reliable rotor faults diagnostics in a real time or for continual IM monitoring.

Author details

Ivan Jaksch

Technical university of Liberec, Czech Republic

Acknowledgement

The author gratefully acknowledges the contributions of the Grant Agency of Czech Republic.

Appendix

Nomenclature

AM	Amplitude Modulation
PM	Phase Modulation
JAPM	Joint Amplitude Phase Modulation
MCDA	Motor Current Demodulation Analysis
MCS	Motor Current Signature -a_{APL}, a_{APH}
IM	Induction motor
a_{AL}, a_{AH}	Low and high spectral magnitudes of AM
a_{PL}, a_{PH}	Low and high spectral magnitudes of PM
a_{APL}, a_{APH}	Low and high spec. mag. of MCS at JAPM –real motor state
f_{sp}	Slip pole frequency – $2sf_i$
f_r	Motor rotation frequency
f_i	Motor supply frequency
f_{load}	Time varying load frequency
$f_{L,H}$	Low and high spectral sidebands frequencies
$H(t)$	Analytical signal created by Hilbert transform
I_i	Induction motor current amplitude
I_{nom}	Induction motor nominal current amplitude
$i_d(t)$, $i_q(t)$	Real and imaginary part of space vector
I_{spa}	AM current amplitude at f_{sp} frequency, **broken bars fault indicator**
I_{spp}	PM current amplitude at f_{sp} frequency

I_{ra} AM current amplitude at f_r frequency, dynamic eccentricity fault indicator

I_{rp} PM current amplitude at f_r frequency

I_{load} AM current amplitude at f_{load} frequency,

$J_i(I_{spp})$ i^{th}-order Bessel function of the first kind

$k_{sp,}$ Fault severity coefficient for broken bars

k_r Fault severity coefficient for dynamic eccentricity

$magH(t)$ Absolute value of the complex analytical signal

n_{rb} Number of rotor bars

p Number of motor poles

$P(t)$ Analytical signal from the space vector

s Per unit slip

$\beta(t)$ Phase of the complex analytical signal

φ Phase angle of PM in reference to AM at broken bars

8. References

Bellini, A., Filippeti, F., Tassoni, C. & Kliman, G.B. (2001). Quantitative evaluation of induction motor broken bars by means of electrical signature analysis. *IEEE Transactions on industry applications*, vol. 37 pp. 1248-1255.

Bellini, A., Filippetti, F., Tassoni, C. & Capolino, G.-A. (2008) Advances in Diagnostic Techniques for Induction Machines, *IEEE Trans. on Industrial. Electronics*, vol. 55, pp 4109-4126.

Bendat, J.S. (1989). *The Hilbert Transform*, Bruel &Kjaer Publication BT0008-11, DK 2850, Naerum, Denmark.

Blödt, M., Bonacci, D., Regnier, J., Chabert, M. & Faucher, J. (2008). On-line monitoring of mechanical faults in variable-speed induction motor drives using Wigner distribution, *IEEE Transactions on Industrial Electronics.*, vol. 55, no. 2, pp. 522–533

Concari, C., Franceschini, G. & Tassoni, C. (2008). Differential Diagnosis Based on Multivariable Monitoring to Assess Induction Machine Rotor Conditions, *IEEE Transactions on Industrial Electronics*, vol. 55, no. 12, pp. 4156-4166.

Cusido, J., Romeral, J.A., Ortega, J.A., Rosero, A. & Garcia Espinosa A. (2008). Fault Detection in Induction Machines Using Power Spectral Density in Wavelet Decomposition, *IEEE Transactions on Industrial Electronics,vol.55*, no.2, pp.633-643

Dorrell, D. G., Thomson, W. T. & Roach, S. (1997). Analysis of airgap flux, current, and vibration signals as a function of the combination of static and dynamic airgap eccentricity in 3-phase induction motors, *IEEE Transactions on Industrial Application.*, vol. 33, no. 1, pp. 24–34.

Drif, M. & Cardoso, A.J.M. (2008). Airgap-Eccentricity Fault Diagnosis, in Three-Phase Induction Motors, by the Complex Apparent Power Signature Analysis, *IEEE Trans. on Industrial Electronics*, vol. 55, no. 3, pp. 1404-1410, March 2008.

Jaksch, I. (2003). Faults diagnosis of three-phase induction motors using envelope analysis, *Proceedings of 4th IEEE International Symposium on Diagnostics for Electric Machines, Power Electronics and Drives*, ISBN 0-7803-7838-5, pp. 289–295, Atlanta, USA, 9/2003

Jaksch, I. & Zalud, J. (2010). Rotor Fault Detection of Induction Motors by Sensorless Irregularity Revolution Analysis, *Proceedings of XIX International Conference on Electric Machines*, ISBN 978-1-4244-4175-4, Paper number RF-001511, Rome, 9/2010.

Joksimovic, G. M. (2005) Dynamic simulation of cage induction machine with air gap eccentricity, *Proceedings of Institute. Elect. Eng.—Electric Power Appl.*, vol. 152, no. 4, pp. 803–811.

Jung, J.H., Lee, J.J. &. Kwon B. H. (2006). Online Diagnosis of Induction Motors Using MCSA. *Transactions on Industrial Electronics*, vol.53, no.6, pp.1842-1852..

Kia, S.H., Henao, H. & Capolino, G.-A. (2009). Diagnosis of broken bar fault in induction machines using discrete wavelet transform without slip estimation, *IEEE Transactions on Industry Applications*, vol. 45, no. 4, pp. 1395-1404.

Kia, S.H.; Henao, H.& Capolino, G. (2010). Torsional vibration Assesment Using Induction Machine Electromagnetic Torque Oscillation, *IEEE Trans. on Ind. electronics*, vol. 57, pp. 209-219.

Kral, C., Pirker, F., Pascoli, G. & Kapeller, H. (2008). Robust Rotor Fault Detection by Means of the Vienna Monitoring Method and a Parameter Tracking Technique, *IEEE Transactions on industrial electronics*, vol. 55, pp. 4229-4237

Lebaroud, G. & Clerc, A. (2008). Classification of Induction Machine Faults by Optimal Time–Frequency Representations, *IEEE Trans. on Industrial Electronics*, vol. 55, no. 12, pp. 4290-4298.

Mirafzal,B. & Demerdash, N. (2005). Effects of Load Magnitude on Diagnosing Broken Bar Faults in Induction Motors Using the Pendulous Oscillation of the Rotor Magnetic Field Orientation, *IEEE Transactions on industry applications*, vol. 41, pp. 771-783.

Nandi, S., Toliyat, H. A. & Li, X. (2005). Condition monitoring and fault diagnosis of electrical motors - A review, *IEEE Transactions on Energy Conversion.*, vol. 20, no. 4, pp. 719–729.

Randall, R.B. (1987). *Frequency analysis*, Bruel&Kjaer Publication, ISBN 87 8735 14 0

Riera-Guasp, M., Antonino-Daviu J. A., Roger-Folch J. & Molina Palomares, M. P. (2008). The use of the wavelet approximation signal as a tool for the diagnosis of rotor bar failures, *IEEE Trans. Industrial Appl.*, vol. 44, no. 3, pp. 716–726.

Su, H. &. Chong, K. T. (2007). Induction machine condition monitoring using neural network modeling, *IEEE Transactions on Ind. Electronics*, vol. 54, no. 1, pp. 241–249.

Thomson, W. & Fenger, M. (2001). Current signature analysis to detect induction motor faults. *IEEE Industry Application Magazine.*, vol. 7, no. 4, pp. 26–34.

Zhang, P, Du, Y., Habetler, T. G. & Lu B. (2011). A Survey of Condition Monitoring and Protection Methods for Medium-Voltage Induction Motors, *Transactions on Industry application*, vol.47, pp.34-46

Zidani, F., Diallo, D., Benbouzid, M.E.H. & Nait-Said, R. (2008). A Fuzzy-Based Approach for the Diagnosis of Fault Modes in a Voltage-Fed PWM Inverter Induction Motor Drive, , *IEEE Transactions on Industrial Electronics*, vol.55, no.2, pp.586-593

Chapter 10

Predictive Maintenance by Electrical Signature Analysis to Induction Motors

Erik Leandro Bonaldi, Levy Ely de Lacerda de Oliveira,
Jonas Guedes Borges da Silva, Germano Lambert-Torresm
and Luiz Eduardo Borges da Silva

Additional information is available at the end of the chapter

1. Introduction

Industries always try to increase the reliability of their productive process. In this context, predictive maintenance performs a fundamental role in order to reach high availability and reliability concerning their pieces of equipment. Predictive maintenance can be understood as the action on the equipment, system or installations based on the previous knowledge about the operation condition or performance, obtained by means of parameters previously determined (Bonaldi et al, 2007).

Since the induction motors are the center of the vast majority of the industrial processes, this chapter gives total emphasis to the failure analysis and identification of this kind of electrical machine. Like all the rotating machines, the induction motors are exposed to many different adversities such as thermal and environmental stresses and mechanical damages, which demand maximum attention (Lambert-Torres et al., 2003). Usually, in industries, attention must be even larger since the downtime costs are very high. High and medium voltage induction motors are highly used in industrial processes. Many of them are strategic to the productive process and, because of that, looking for solutions that minimize the failure statistics is mandatory. In most cases, these motors are highly reliable and extremely expensive, forcing the company to operate without a stand-by.

Many predictive techniques are applied to these motors to reduce the number of unplanned outage. The most common techniques applied to fault detection in induction motors are: vibration analysis, acoustical analysis, speed oscillations, partial discharges, circuit analysis, etc. The analyses based on mechanical concepts are established, but the techniques based on electrical signature analysis are being introduced only now. Because of that status, the application of Electrical Signature Analysis (ESA) to industries is the concern of this chapter.

The industries currently look for products and outside services for predictive maintenance. In many cases, the outside service company or even the industrial plant predictive group make mistakes that can compromise the whole condition monitoring and failure diagnosis process.

In this increasing demand for prediction technology, a specific technique referred as Electrical Signature Analysis (ESA) is calling more and more attention of industries. Considering this context, this chapter intends to disseminate important concepts to guide companies that have their own predictive group or want to hire consultants or specialized service to obtain good results through general predictive maintenance practices and, especially through electrical signature analysis.

Figure 1 presents the comparative between vibration analysis and ESA (considering Motor Current Signature Analysis (MCSA), Extended Park's Vector Approach (EPVA) and Instantaneous Power Signature Analysis (IPSA)), showing which technique is more recommended to a specific kind of problem in a determined part of the rotating drive train. One can say that those techniques are complementary.

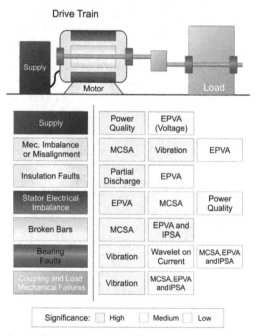

Figure 1. Comparison of predictive maintenance techniques

The main objective of this chapter is to present a procedure to acquire and analyze electrical signals for condition monitoring of electrical machines through motor current signature analysis in order to get the best possible results in an industrial environment. As secondary contributions, the chapter intends to disseminate important concepts to guide companies that have their own predictive group or want to hire consultants or specialized service to obtain good results through general predictive maintenance practices and, especially through electrical signature analysis. For this purpose, the chapter presents a discussion between condition monitoring and troubleshooting, pointing the differences between both approaches and the main benefits and problems involved with each one.

The result of the proposed discussion in this chapter is a procedure of acquisition and analysis, which is presented at the end of the chapter and intends to be a reference to be used by industries that have a plan to have ESA as a monitoring condition tool for electrical machines.

2. Considerations about maintenance

The motors are the center of the majority of the industrial production processes. Therefore, these machines deserve concerns to increase the reliability of the productive process. In this sense, many techniques have been developed for an on-line motor monitoring of the behavior and performance.

Monitoring condition of electric machines is an evaluation continuous process of the health of equipment during all its useful life. The main function of a monitoring predictive system is to recognize the development of failures in an initial state. For the maintenance department, each failure must be detected as soon as possible in order to promote a programmed stop of the machine.

The process of continuous monitoring of the condition of vital electric machines for the production process brings significant benefits for the company. The main benefits are: bigger efficiency of the productive process, reduction of the losses for not-programmed stops, increase of the useful life of the equipment, and build a historical of failure (Legowski et al., 1996; Tavner et al., 1997; Thomson & Fenger, 2001).

A continuous monitoring system must observe parameters that give to the maintenance team trustworthy information for the decision-making. The more usual monitored parameters are: voltage and current of the stator; temperature of the nucleus; level of vibration; instantaneous power; level of contamination in the lubricant of the rolling; speed of rotation; flow of escape; and so on.

In such a way, it can be noticed that this area of the technology demands knowledge of the functioning of electric machines, instrumentation, microprocessors, processing of signal, analysis of materials, chemical analysis, analysis of vibrations, etc.

2.1. Classification of the maintenance activities

"Maintenance" can be understood as the action to repair or to execute services in equipment and systems. It can have its activities classified in four main groups:

a. Corrective maintenance: this is the most primary form of maintenance. It occurs after a failure carried out. Usually, it becomes the unavailable equipment for use. Many disadvantages of this type of maintenance are clear. As examples, the systematic occurrence of not-programmed stops, lesser time of useful life for the machine, bigger consumption of energy (since with the presence of the failure the motor needs more current keeping the constant torque) can be cited.

b. Preventive maintenance: this is the name that receives a set of actions developed with the intention of preventing the occurrence of unsatisfactory conditions, and consequently, reducing the number of corrective actions. When preventive maintenance plan is elaborated, a set of technical measurements must be created in order to increase the machine reliability and decrease the total cost of the maintenance. A preventive maintenance program can still choose for one of the three types of activities: continuous monitoring; periodic measurements; or predictive techniques.

c. Predictive maintenance: as it can see previously, the predictive maintenance can be a sub-area of the preventive maintenance. However, the predictive maintenance presents some proper characteristics as:

- Support in not invasive techniques, that is, it is not necessary to stop the operation of the machine for its application
- Elimination of corrective maintenance;
- Not consideration of information as the durability of components;
- On-line or off-line can be effected through techniques.

d. Systematic maintenance: characterized for the substitution of components of the equipment or for the substitution of the equipment as a whole (Bonaldi et al., 2007).

2.2. Status of predictive maintenance

Usually, industries have the vast majority of their condition monitoring programs based on the mechanical parameters analysis. The most common methods applied are: Vibration Analysis, Acoustical Analysis, Shock Pulse and Speed Fluctuations. Other techniques involving mechanical concepts are also applied such as temperature monitoring, oil and gases analysis, etc.

When involving electrical concepts, intrusive methods are more common used in industries such as surge test, polarization index, hipot tester, motor circuit analysis (MCA), etc. These techniques are more correctly classified as preventive maintenance, being performed at planned outages.

Concerning motor condition monitoring through non-intrusive electrical methods in Brazil, one can observe more often the RMS voltage and current monitoring. For example, broken bars produce current oscillations that can be observed through an ampere-meter installed in the electric panel. But, again, it is not possible to separate load oscillations from broken bars.

This way, a more reliable program to detect electrical and mechanical problems must consider the introduction of new condition monitoring tools, mainly those related to electrical signature that has been neglected until now. Since the petrochemical industry constantly aims to increase the process reliability and operational continuity, a very interesting and little explored field surfaces, which is the introduction of predictive maintenance techniques based on electrical signature analysis.

3. Common failures in three-phase induction motors

Consider the following brief description of the most common failures that can be avoided through the adoption of condition monitoring methods:

a. Bearings Faults: can be caused by incorrect lubrication, mechanical stresses, incorrect assembling, misalignment, etc. They can affect all the bearing parts such as inner and outer races, cage and balls or rolls.

b. Stator Winding Faults: normally a consequence of overheating, contamination, project errors, etc., possibly causing shorted turns, shorted coils (same phase), phase to phase, phase or coil to ground and single phasing. Such failures cause stator electrical imbalance as well as variations in the current harmonic content. Mechanical problems can also occur in the stator such as loosen edges, but this is statistically less frequent.

c. Rotor Faults: usually caused by broken bars or broke end rings, rotor misalignment and imbalance.

Faults in the coupling (pulley, belt and gear mesh) and in the attached load also can be diagnosed. The failures are also related to the petrochemical process different characteristics. For example, at off-shore plants, the motors start directly from the mains. This demands high start currents and causes pulsating torques which contributes to the origin of rotor and stator faults. Furthermore, outdoors motors present more incidence of failure than indoor motors. The same statistic holds for high voltage motors and high speed motors when compared with low voltage motors and low speed motors.

3.1. Abnormalities in three-phase induction motors

The main focus of problems in three-phase induction motors are in the stator and the supports. The main causes of failures are: superheating, imperfections in the isolation, mechanical bearings and electric failures. Figure 2a presents a division of the failures in three-phase induction motors with squirrel steamer and power of 100 HP or higher (Bonnett & Soukup, 1992).

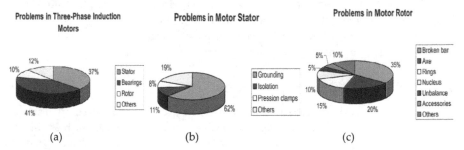

Figure 2. Problems in: (a) Three-Phase Induction Motors, (b) Motor Stator, and (c) Motor Rotor.

In one hand, the main source of electrical problems in induction motor is in stator that totalizes 37% of the total of failures. Figure 2b details different type of problems in the motor stators. In the other hand, problems in the motor rotor totalize 10% of the total of motor failures, and they are shown in Figure 2c.

3.2. Relation between motor specification and failure mechanism

Many failures can be deriving of incorrect specifications. The specification of a motor must consider the mechanical and electric conditions, and the environment in which the machine goes to work. The monitored parameters are affected by these operational conditions. In terms of the mechanical conditions, the failures appear as resulted of the behavior of the load. Amongst the main problems they are distinguished:

- Successive overloads that can cause superheating and/or damages to the bearing;
- Pulsing load that can cause damages to the bearing;
- Repeated departures that can damage the machine bearing;
- Vibration that can be transmitted to the machine causing damages to the bearing.

In terms of the electric conditions, the failures can result of the electrical power system characteristics or the load feeder by the motor. Amongst the main problems they are distinguished:

- Slow fluctuations of voltage being able to cause loss of stop power and of the machine.
- Brusque fluctuations of voltage being able to cause failure in the isolation.

In terms of the environment conditions, the failures can result of the characteristics of the process in which the machine is being used. Amongst the main problems they are distinguished:

- High temperatures that can cause the deterioration of isolation.
- Humidity and pollution that can respectively cause imperfections and contamination of the isolation.

Thus, it is clear that the failures that occur in electric machines depend on the type of machine and the environment where it is working. What it is really important to observe it is that the failure mechanism happens in gradual way, from an initial defect up to real failure. The time of propagation of the failure depends on some factors. However, the major parts of the failures present initial pointers of its presences and are exactly in these initial indications that the predictive maintenance must act (Bonaldi et al., 2003).

4. Electrical signature analysis overview

Electrical Signature Analysis (ESA) is the general term for a set of electrical machine condition monitoring techniques through the analysis of electrical signals such as current and voltage. These techniques are: Current Signature Analysis (CSA), Voltage Signature Analysis (VSA), Extended Park's Vector Approach (EPVA), Instantaneous Power Signature

Analysis (IPSA), among others. The electrical motor of the rotating system under analysis is analyzed for the failure diagnosis purposes, acting as a transducer in this process. Variations in the voltage and current signals are analyzed in relation to some failure patterns.

The industrial application of ESA techniques aims to improve the equipment reliability once those techniques imply greater robustness to the diagnosis. The expected results are: downtime reduction, increase in the machine availability, maintenance costs reduction, better management and planning of maintenance, etc.

The inherent benefits in ESA are: non-intrusive; it does not demand sensors installed in the rotating drive train; it is not necessary to be suited for classified areas (the sensors can be installed in the motor control centre (MCC) free of explosive mixtures); it presents high capability of remote monitoring, reducing the human exposure to risks; it can be applied to any induction motor without power restriction; it presents sensitivity to detect mechanical failures in the motor and load, electrical failures in the stator and problems in the mains, etc.

For these reasons, one recommends the application of these techniques (together with the mechanical approaches) in order to prevent catastrophic failures; improve the safety and the reliability of the productive process; reduce the downtime, improve the condition monitoring of motors installed in places of difficult access and improve the motor management in the maintenance context for reliability purposes.

Among the several ESA techniques, two of them are considered in this chapter: MCSA and EPVA.

The stator line current spectral analysis has been widely used recently for the purpose of diagnosing problems in induction machines. This technique is known as Motor Current Signature Analysis (MCSA) and the current signal can be easily acquired from one phase of the motor supply without interruption of the machine operation. In MCSA the current signal is processed in order to obtain the frequency spectrum usually referred to as current signature. By means of the motor signature, one can identify the magnitude and frequency of each individual component that constitutes the signal of the motor. This characteristic permits identifying patterns in the signature in order to differentiate healthy motors from unhealthy ones and point where the failures happen. Although it is important to say that the diagnosis is something extremely complicated, e.g., the decision of stopping or not the productive process based on the current spectrum indications is always not elementary and demands experience and knowledge of the process.

4.1. Current and voltage signature analysis

CSA – Current Signature Analysis or VSA – Voltage Signature Analysis techniques are used to generate analyses and trend of electric machines dynamically. They aim to detect predictive problems in a rotating electric machine, such as: problems in the stator winding, rotor problems, problems on the engagement, problems in bound load, efficiency and system load; problems in the bearing, among others. It may initially cause a certain astonishment that the electrical signals contain information in addition to the electrical characteristics of the machine under supervision, but they work for mechanical defects as a transducer, allowing the electrical signals (voltage and/or current) can carry information of electrical and mechanical problems until the power panel of the machine.

The signs of current and/or voltage of one or three phases of the machine produce, after analyzed, the *signature of machine*, i.e., its operating pattern. This signature is composed of magnitudes of frequencies of each individual component extracted from their signals of current or voltage. This isolated fact itself is an advantage, as it allows the monitoring of the evolution of the magnitudes of the frequencies, which can denote some sort of evolution of operating conditions of the machinery. The response that the user of such a system needs to know is whether your machine is "healthy" or not, and that part of the machine the failure might occur.

This analysis (diagnosis) is not something easy to be done, because it involves a set of comparisons with previously stored patterns and own "history" of the machine under analysis. In this instant, normally a specialist is called to produce the final diagnosis, generating the command when stopping the machine.

4.2. Motor Current Signature Analysis (MCSA)

MCSA is the technique used to analyze and monitor the trend of dynamic energized systems. The appropriate analysis of the results of applying predictive technique helps in identifying problems in stator winding, rotor problems, problems in the coupling, problems in attached load, efficiency and system load, problems in the bearing, among others.

This technique uses the induction motor as a transducer, allowing the user to evaluate the electrical and mechanical condition from the panel and consists primarily in monitoring of one of the three phases of the supply current of the motor. A simple and sufficient system for the implementation of the technique is presented in the Figure 3a.

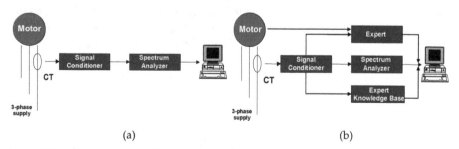

(a)   (b)

Figure 3. Basic System for Spectral Analysis of the Current

Thus, the current signal of one of the phases of the motor is analyzed to produce the power spectrum, usually referred to as <u>motor signature</u>. The goal is to get this signature to identify the magnitude and frequency of each individual component that integrates the motor current signal. This allows that patterns in current signature be identified to differentiate "healthy" motors from "unhealthy" ones and even detect in which part of machine failure should occur.

However, it is important to note that the diagnosis is something extremely complicated, i.e. the definition of stopping or not the production process in view of the indications of the power spectrum is always difficult and requires experience and knowledge of the process. This time, it is important to consider the expert knowledge and the data history of the behavior of the set (motor, transmission system and load). For this reason, an automatic diagnostic system that combines the data history of the motor to the attention of specialist is a niche market quite promising. This way, the automatic diagnosis and analysis system is no longer as simple as the model shown in Figure 3a and can be represented by the new elements in Figure 3b.

The Fast Fourier Transform (FFT) is the main tool employed, however some systems employ in conjunction with other techniques to increase the ability of fault detection since signal acquisition, through processing, up to the diagnostic step. Among the most important issues related to acquisition of signals and the FFT include:

a. **Frequency range:** the frequency response is typically required in MCSA 5 kHz. This way, the bandwith of the transducers used must be at least 10 kHz.
b. **Nyquist theorem:** this theorem states that for any signal to be reconstructed without significant losses must be removed samples with twice the maximum frequency of the signal. In practice it uses 10 times the maximum frequency and ensures excellent accuracy.
c. **Resolution:** spectral lines resolution, i.e. the distance between two spectral is given by (1):

$$\Delta f = \frac{f_s}{N} \tag{1}$$

Where Δf is the spectral resolution, f_s is the sampling frequency used, and N is the number of samples.

Other important issues are related to the own operation of induction motors. The first one is the induction motor synchronous speed that is given by (2):

$$N_S = \frac{f_1}{p} \tag{2}$$

Where f_1 represents the power frequency, N_s is the velocity of the rotating field, and p is the number of motor pole pairs.

From the synchronous speed, two important concepts for the current signature analysis can be presented: the slip speed and the slip. In MCSA is important to note that the rotor speed is always less than the synchronous speed. The frequency of the induced currents in the rotor is a function of frequency and power slip. When operating without load, the rotor rotates at a speed close to the synchronous speed. In this case, torque should be just sufficient to overcome friction and ventilation. The difference between the rotor speed (Nr) and the synchronous speed (Ns) is named as slip speed (N_{Slip}):

$$N_{Slip} = N_s - N_r \tag{3}$$

When mechanical load is attached to the rotor demanding torque the rotor speed decreases. In this turn, the slip speed increases and also the current in the rotor to provide more torque. As the load increases, the rotor continues having reduced its speed relative to synchronous speed. This phenomenon is known as motor slip, denoted by s.

$$s = \frac{\left(N_s - N_r\right)}{N_s} \tag{4}$$

Another important definition refers to slip frequency. The frequency induced in the rotor is correctly set to slip frequency and is given by:

$$f_2 = \left(N_s - N_r\right) \cdot p \tag{5}$$

As noted, the rotor frequency is directly proportional to the slip speed and the number of pair of poles. Thus:

$$s \cdot N_S = N_S - N_r \text{ and } p \cdot N_S = f_1 \text{ then } f_2 = s \cdot f_1 \tag{6}$$

This is a very important result for MCSA once the current frequency is rotor slip function. The characteristic frequencies are well known. The patterns of these failures are presented below.

The stator line current spectral analysis has been widely used recently for the purpose of diagnosing problems in induction machine. This technique is known as MCSA and the current signal can be easily acquired from one phase of the motor supply without interruption of the machine operation. In MCSA the current signal is processed in order to obtain the frequency spectrum usually referred to current signature. By means of the motor signature, one can identify the magnitude and frequency of each individual component that constitutes the signal of the motor. This characteristic allows identifying patterns in the signature in order to differentiate healthy motors from unhealthy ones. Mechanical failures such as rotor imbalance, shaft misalignment, broken bars and bearing problems are common in induction machines applications and commonly discussed or presented when talking about MCSA. Another very important cause of poor functioning of induction motor is load mechanical failure. When a mechanical failure is present either in the motor, or in the transmission system or in the attached load, the frequency spectrum of the line current, in other words, the motor signature, becomes different from that of a non-faulted machine.

When a mechanical failure occurs in the attached load of an induction motor, multiples rotational frequencies appear in the stator current due to the load torque oscillation (Benbouzid, 2000). These frequencies are related to the constructive characteristics of the load and the transmission system, and an abnormal value of a given frequency expresses a specific failure, and more, the severity of this failure. The frequency component that appears in the stator current spectrum can be expressed by:

$$f_{lf} = f_1 \pm \kappa f_r \tag{7}$$

Where f_{lf} is the characteristic frequency of the load fault, f_1 is the supply frequency, κ is the constant resulting from the drive train constructive characteristics and f_r is the motor rotational frequency.

It is known that when a mechanical failure has developed in the load, it generates an additional torque (T_{lf}). Thus, the overall load torque (T_{load}) can be represented by an invariable component (T_{const}) plus this additional variable component which varies periodically at a characteristic frequency ω_{lf} in (8)

$$T_{load}(t) = T_{const} + T_{lf} \cos(\omega_{lf} t) \tag{8}$$

Where T_{lf} is the amplitude of the load torque oscillation caused by the load mechanical failure and $\omega_{lf} = 2\pi f_{lf}$. Also, the torque relates to the rotational frequency (ω) can be expressed by:

$$T(t) = T_{motor}(t) - T_{load}(t) = J\frac{d\omega_r(t)}{dt} \tag{9}$$

Where J is the total inertia of machine and load. Thus:

$$J\frac{d\omega_r(t)}{dt} = T_{motor}(t) - T_{const} - T_{lf} \cos(\omega_{lf} t) \tag{10}$$

In steady state, $T_{motor} = T_{const}$ and:

$$\frac{d\omega_r(t)}{dt} = -\frac{1}{J}\left(T_{lf}\cos(\omega_{lf}t)\right) \text{and} \ \omega_r(t) = -\frac{T_{lf}}{J}\int\cos(\omega_{lf})dt + Const. \tag{11}$$

Then

$$\omega_r(t) = -\frac{T_{lf}}{J\omega_{lf}}\sin(\omega_{lf}t) + \omega_{r0} \tag{12}$$

Observing (12), the mechanical speed consists of a constant component ω_0 and a component which varies according to a sinusoidal signal. Then, the integration of mechanical speed results in the mechanical rotor position $\theta(t)$:

$$\theta_r(t) = \frac{T_{lf}}{J\omega_{lf}^2}\cos(\omega_{lf}t) + \omega_{r0}t \tag{13}$$

The rotor position oscillations act on the magneto motive force (MMF). In normal conditions, the MMF referred to as the rotor ($F_r{}^{(R)}$) can be expressed by (14).

$$F_r^{(R)}(\theta',t) = F_r \cos(p\theta' - s\omega_1 t) \tag{14}$$

Where θ' is the mechanical angle in the rotor reference frame, p is the number of pole pairs, ω_1 is the synchronous speed, s is the motor slip, and F_r is the rotor MMF.

Figure 4 shows a phasorial diagram for the rotor MMF (R axes) referred to the stator frame (S axes), the difference can be expressed by the angle θ'.

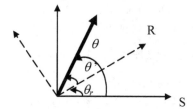

Figure 4. Phasorial diagram of the rotor MMF referred to the stator frame

According the Figure 4 and Equation (14) and replacing (13) in (15), it results in:

$$\theta = \theta' + \theta_r \text{ and } F_r(\theta,t) = F_r \cos[p(\theta - \theta_r) - s\omega_1 t] \tag{15}$$

$$F_r(\theta,t) = F_r \cos\left(p\theta - p\omega_{r0}t - \frac{pT_{lf}}{J\omega_{lf}^2}\cos(\omega_{lf}t) - s\omega_1 t\right) \tag{16}$$

Doing $\beta = pT_{lf}/J\omega_{lf}^2$ and using the relation $\omega_{r0} = (1-s)\omega_1/p$, it produces:

$$F_r(\theta,t) = F_r \cos\left(p\theta - \omega_1 t - \beta\cos(\omega_{lf}t)\right) \tag{17}$$

Where β is the modulation index and generally $\beta \ll 1$.

At this point, it is important to notice that the term $\beta\cos(\omega_{lf}t)$ means a phase modulation. The failure does not have direct effect on stator MMF which can be expressed by:

$$F_s(\theta,t) = F_s \cos(p\theta - \omega_1 t - \varphi_s) \tag{18}$$

Where φ_s is the initial phase between rotor and stator MMFs.

Supposing for the sake of simplicity the value of the air gap permeance Λ constant (because slotting effects and eccentricity were neglected), the air gap flux density B can be expressed by the product of total MMF and Λ:

$$B(\theta,t) = \left[F_s(\theta,t) + F_r(\theta,t) \right] \Lambda \text{ and}$$

$$B(\theta,t) = B_s \cos\left(p\theta - \omega_1 t - \varphi_s\right) + B_r \cos\left(p\theta - \omega_1 t - \beta \cos\left(\omega_{lf} t\right)\right) \tag{19}$$

As the flux $\varphi(t)$ is obtained by the integration of the flux density $B(\theta,t)$, then all phase modulation existing in the flux density also exists in the flux $\varphi(t)$. It is important to explain that the winding structure affects only the flux amplitude and not its frequencies. Thus:

$$\varphi(t) = \varphi_s \cos\left(\omega_1 t - \varphi_s\right) + \varphi_r \cos\left(\omega_1 t - \beta \cos\left(\omega_{lf} t\right)\right) \tag{20}$$

The relationship between the flux and the current is given by the equation (21).

$$V(t) = R_s I(t) + \frac{d\varphi(t)}{dt} \tag{21}$$

Where R_s is the stator resistance. Thus,

$$I(t) = \frac{V(t)}{R_s} - \frac{1}{R_s}\frac{d\varphi(t)}{dt} \tag{22}$$

And as:

$$\begin{aligned}\frac{d\varphi(t)}{dt} &= -\omega_1 \varphi_s \sin\left(\omega_1 t + \varphi_s\right) - \omega_1 \varphi_r \sin\left(\omega_1 t + \beta \cos\left(\omega_{lf} t\right)\right) \\ &+ \omega_{lf} \beta \varphi_r \sin\left(\omega_1 t + \beta \cos\left(\omega_{lf} t\right)\right) \sin\left(\omega_{lf} t\right)\end{aligned} \tag{23}$$

With the last term being neglected once $\beta \ll 1$. Finally:

$$I(t) = \underbrace{\frac{V(t)}{R_s} + \frac{1}{R_s}\omega_1 \varphi_s \sin\left(\omega_1 t + \varphi_s\right)}_{\text{Stator}} + \underbrace{\frac{1}{R_s}\omega_1 \varphi_r \sin\left(\omega_1 t + \beta \cos\left(\omega_{lf} t\right)\right)}_{\text{Rotor}} \tag{24}$$

$$I(t) = \underbrace{I_{st} \sin\left(\omega_1 t + \varphi_s\right)}_{i_{st}} + \underbrace{I_{rt} \sin\left(\omega_1 t + \beta \cos\left(\omega_{lf} t\right)\right)}_{i_{rt}} \tag{25}$$

Notice that the term i_{st} results from stator MMF and it is not influenced by the torque oscillation, and the term i_{rt} results from the rotor MMF and presents phase modulation due to torque oscillations. And also, when the motor is healthy β is null.

Considering the component i_{rt} with phase modulation in (14) given in its complex form:

$$i_{rt}(t) = I_{rt} e^{j\left(\omega_1 t + \beta \cos\left(\omega_{lf} t\right)\right)} \tag{26}$$

Applying a Discrete Fourier Transform (DFT) in (26), as well known from communications theory, it can be expressed by (27).

$$I_{rt}(f) = I_{rt} \sum_{n=-\infty}^{\infty} j^n J_n(\beta) \delta\left(f - \left(f_1 + nf_{lf}\right)\right)$$

(27)

Where J_n denotes the nth-order Bessel function of first kind and $\delta(f)$ is the Dirac delta function. Since β is so small, the Bessel functions of order $n \geq 2$ can be neglected.

Finally, the Power Spectral Density (PSD) of the stator current, considering the approximations used, is given by:

$$\left|I(f)\right| = \left(I_{st} + I_{rt}J_0(\beta)\right)\delta\left(f - f_1\right) + I_{rt}J_1(\beta)\delta f - \left(f_1 \pm f_{lf}\right)$$

(28)

It is clear that the phase modulation leads to sideband components of the fundamental at $f_1 \pm f_{lf}$ as it happens in an amplitude modulation. Considering all the development accomplished in this section and the result in (28), the load failure patterns can be presented.

4.3. Voltage Signature Analysis (VSA)

The technique of Voltage Signature Analysis follows the same strategy of analysis of the current signature; however the signal is analyzed from the voltage supply of the motor. This technique is most often used in analysis of generating units. In the case of motors, it can be usefully employed in cases of problems from the motor power and the analysis of electric stator imbalance in conjunction with the analysis of the current signature. It can be used also to know the origin of certain components in the power spectrum, that is, it can be used to infer if the source of the component comes from the mains or has its origin in the array itself.

4.4. Instantaneous Power Signature Analysis (IPSA)

The analysis of the instantaneous power is another failure analysis technique based on spectral analysis. The big difference between this technique and MCSA and VSA is that it considers the information present in voltage and current signals of a motor phase concurrently and demodulated fault component appears under the name of Characteristic Frequency. Considering an ideal three phase system, instant power $p(t)$ is given by:

$$p(t) = v_{LL}(t)i_L(t)$$

(29)

Where v_{LL} is the voltage between two terminals of the motor and i_L is the current entering one of these terminals. And, a motor under normal conditions, i.e. without breakdowns, and constant velocity, one has:

$$v_{LL}(t) = \sqrt{2}V_{LL}\cos(\omega t)$$

(30)

$$i_{L,0}(t) = \sqrt{2}I_L \cos\left(\omega t - \varphi - \frac{\pi}{6}\right) \tag{31}$$

$$p_0 = v_{LL}(t)i_{L,0}(t) = V_{LL}I_L\left[\cos\left(2\omega t - \varphi - \frac{\pi}{6}\right) + \cos\left(\varphi + \frac{\pi}{6}\right)\right] \tag{32}$$

Where V_{LL} and I_L are the RMS values of voltage and current line, ω is the angular frequency and φ is the phase angle of the motor load.

Let's consider now the presence of a mechanical fault in the drive train, resulting in the appearance of motor torque oscillations accompanied by surges of speed and slip, which in turn result in modulations in the current spectrum.

For simplicity, it is considered that the failure cause only an amplitude modulation on the stream of the stator by deleting the effect on stage. It could also prove that phase modulations, in function of torque oscillations, appear in current as amplitude modulations by processed result from similar functions to the Bessel functions. The modulated current i_L can be expressed by:

$$i_L = i_{L,0}(t)\left[1 + M\cos\left(\omega_f t\right)\right]$$
$$= i_{L,0}(t) + \frac{MI_L}{\sqrt{2}}\left\{\cos\left[\left(\omega + \omega_f\right)t - \varphi - \frac{\pi}{6}\right] + \cos\left[\left(\omega + \omega_f\right)t - \varphi - \frac{\pi}{6}\right]\right\} \tag{33}$$

Where M is the index modulation and ω_f is the angular frequency of the failure.

The expression of instant power results in:

$$p(t) = p_0(t) + \frac{MV_{LL}I_L}{2}\left\{\cos\left[\left(2\omega + \omega_f\right)t - \varphi - \frac{\pi}{6}\right] + \cos\left[\left(2\omega - \omega_f\right)t - \varphi - \frac{\pi}{6}\right] + \right.$$
$$\left. +2\cos\left(\varphi + \frac{\varphi}{6}\right)\cos\left(\omega_f t\right)\right\} \tag{34}$$

Besides the fundamental component $2\omega/2\pi$ and the lateral bands in $(2\omega \pm \omega_f)/2\pi$, the spectrum of instantaneous power contains an additional component directly related to the modulation caused by failure. This component is named as **Characteristic Component** and can be used as information for the diagnosis of the condition of the machine.

The following simulation which considers a motor current modulation originated by an alleged mechanical failure whose frequency characteristic is of 15 Hz. Note the Figure 5 that the spectrum of voltage does not have any type of modulation, since the current spectrum has lateral bands apart from 15 Hz fundamental's (located at 60 Hz). The instantaneous power spectrum has the fundamental frequency in 120 Hz with modulations of 15 Hz at 105 and 135 Hz, besides presenting the fault feature component in isolated 15 Hz.

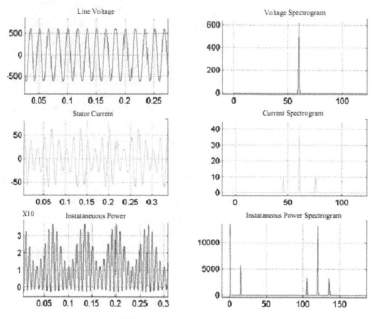

Figure 5. Fault Simulation in 15 Hz and the respective spectra of voltage, current and instant power

4.5. Enhanced Park's Vector Approach (EPVA)

The first research involving the use of Park's vector method for the diagnosis of failures in motors such as short circuit between turns, airgap eccentricity and broken bars, etc.(Cardoso & Saraiva, 1993). At first, the proposed damage detection was based only on the distortion suffered by circle of Park on the emergence and on the aggravation of the damage. More recently, the technique has been improved (now named EPVA) and may be described as following steps. The three phases of currents in a motor can be described by:

$$i_A = i_M \cos(\omega t - \alpha) \tag{35}$$

$$i_B = i_M \cos\left(\omega t - \alpha - \frac{2\pi}{3}\right) \tag{36}$$

$$i_C = i_M \cos\left(\omega t - \alpha + \frac{2\pi}{3}\right) \tag{37}$$

Where i_M is the peak value of the supply current, ω is the angular frequency in rad/s, α the is the initial phase angle in rad, t is the time variable; and i_A, i_B and i_C are respectively the currents in the phases A, B and C. The current components of the Park's vector are given by:

$$i_D = \left(\frac{\sqrt{2}}{\sqrt{3}}\right)i_A - \left(\frac{1}{\sqrt{6}}\right)i_B - \left(\frac{1}{\sqrt{6}}\right)i_C \text{ and } i_Q = \left(\frac{1}{\sqrt{2}}\right)i_B - \left(\frac{1}{\sqrt{2}}\right)i_C \qquad (38)$$

Ideally:

$$i_D = \left(\frac{\sqrt{6}}{2}\right)i_M \cos(\omega t - \alpha) \text{ and } i_Q = \left(\frac{\sqrt{6}}{2}\right)i_M sen(\omega t - \alpha) \qquad (39)$$

Graphically, ideal conditions generate a perfect Park circle centered at the origin of coordinates, as shown in Figure 6.

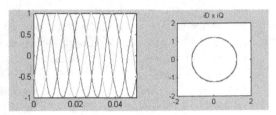

Figure 6. Signals in time and Park circle

Under abnormal conditions of operation, i.e. when the emergence of mechanical or electrical failure, the previous equations are no longer valid and the circle of Park passes to suffer distortions. As these changes in the circle of Park are difficult to be measured, was proposed by EPVA method of observation of spectrum of Park's vector module. The advantage of EPVA technique combines the simplicity of the previous method (analysis of the Park's circle) with spectral analysis capability. In addition, the fundamental component of the motor power is automatically subtracted from the spectrum by Park transformation, causing the failure characteristics components appear prominently. The most important point is the fact that the technique considers the three phases of current, generating a more significant spectrum by encompass information from three phases. This feature is extremely useful in cases where failure can only be detected if considered the three phases. This is the case of unbalanced electric motor fuelled in open loop.

When there is an unbalanced voltage supply, the motor currents can be represented by:

$$i_A = i_d \cos(\omega t - \alpha_d) + i_i \cos(\omega t - \beta_i) \qquad (40)$$

$$i_B = i_d \cos\left(\omega t - \alpha_d - \frac{2\pi}{3}\right) + i_i \cos\left(\omega t - \beta_i + \frac{2\pi}{3}\right) \qquad (41)$$

$$i_C = i_d \cos\left(\omega t - \alpha_d + \frac{2\pi}{3}\right) + i_i \cos\left(\omega t - \beta_i - \frac{2\pi}{3}\right) \qquad (42)$$

Where i_d is the maximum value of the current direct sequence, i_i is the maximum value of reverse sequence current, α_d is the current initial phase angle direct sequence in rad, and β_i is the initial phase angle reverse sequence current in rad. In the Park's vector:

$$i_D = \left(\frac{\sqrt{3}}{\sqrt{2}}\right)\left(i_d \cos\left(\omega t - \alpha_d\right) + i_i \cos\left(\omega t - \beta_i\right)\right) \text{and} \quad i_Q = \left(\frac{\sqrt{3}}{\sqrt{2}}\right)\left(i_d \sin\left(\omega t - \alpha_d\right) - i_i \sin\left(\omega t - \beta_i\right)\right) (43)$$

And the square of the Park's vector module is given by:

$$\left|i_D + j i_Q\right|^2 = \left(\frac{3}{2}\right)\left(i_d^2 + i_i^2\right) + 3 i_d i_i \cos(2\omega t - \alpha_d - \beta_i) \tag{44}$$

Now, just applying the FFT to the square of the Park's vector module and observe that this is composed by a DC level plus one additional term located at twice the supply frequency. It is exactly this additional term that indicates the emergence and intensification of stator electrical asymmetries. Let's the example shown in Figure 7a which is considered an unbalanced feed; and also, the Park circle passes to resemble an ellipse and arises in the spectrum the component located at twice the supply frequency, as shown in Figure 7 b and c. Thus, the whole process can be represented by the elements of Figure 8.

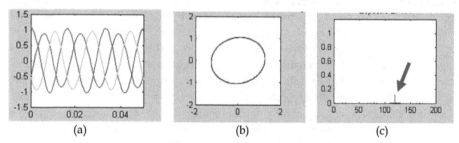

Figure 7. Imbalance between the phases, Park circle distorted and presence of the component at twice the supply frequency

Figure 8. Block Diagram of the EPVA technique

This demonstrates the effectiveness of the component located at twice the supply frequency (in this case 120 Hz) of the EPVA monitoring to diagnosis short circuit between turns. The test procedure was the following: used the Marathon motor failures Simulator Spectra Quest in which *taps* was inserted to the gradual introduction of imbalance in power depending on the insertion of short. Figure 9a presents the characteristics of the motor and the *taps* as to the short are introduced.

Tests have been made in the conditions of non-faulted motor (no imbalance) and five severities of short circuit generating imbalances of 1.2 V, 1.8 V, V, V 5.4 6.7 and 8.5 V. Figure 9b shows the overlap of the spectra of the motor in normal condition (in red) and motor in the worst condition of imbalance (8.5 V) highlighting the component twice the power frequency in the spectrum of Park vector module.

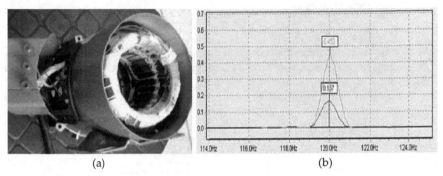

(a)   (b)

Figure 9. Featured for the inserted short-circuit and spectrum of Park vector module

The current trend curve (shown in Figure 10) demonstrates a general growth of electric unbalance component EPVA with increasing the short circuit. Each three points of the curve represent a condition of normal severity, starting and advancing to severity 1 (1.2 V), 2 (1.8 V), 3 (5.4 V), 4 (6.7 v) and 5 (V 8.5). Severity 4 presents amplitude less than Severity 3 due to a change in the equilibrium condition of input voltage shown in the trend curve in tension (shown in Figure 10), being thus possible to separate the effects of those supply imbalances caused by short circuits and other anomalies.

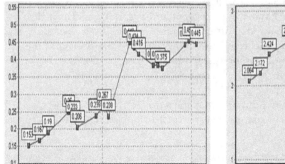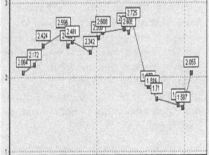

Figure 10. Trend curve to the imbalance component: (a) for current and (b) for voltage

5. Patterns of failures

A fault in any part of the machine is a decrease in this part performance when compared with the minimum requirements specified. Thus the fault results from natural wear, project errors, incorrect installation, poor use or a combination of all of them. If the fault is not identified in time and increases, failure may ensue (Thorsen & Dalva, 1999). Therefore, failure is the reason why the machine breaks down. This way, one tries to identify the fault before it becomes a failure, even when it is incipient.

5.1. Rotor failure patterns

This section shows the failure patterns for rotor problems.

1. Broken Bar: it is the rotor most common problem and the better known pattern. Figure 11 presents this failure pattern, where f is the supply fundamental frequency and s is the motor slip.

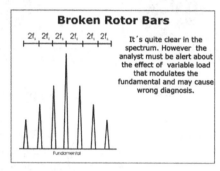

Figure 11. Broken bar pattern

2. Air gap Eccentricity: it is the condition in which the air gap doesn't present a uniform distance between the rotor and stator, resulting in a region of maximum air gap and another region of minimum air gap. There are two kinds of air gap eccentricity: static and dynamic. Figure 12 shows the patterns for both kinds, where f_1 is the supply fundamental frequency, R is number of rotor bars, and CF is the center frequency.
a. Static Eccentricity: the minimal radial air gap position is fixed in the space. The stator core is bowed or there is an incorrect positioning between the rotor and the stator generated as a consequence of misalignment. Besides those possibilities, constructive aspects permit an inherent level of eccentricity due to the tolerances of the manufacturing process.
b. Dynamic Eccentricity: the minimum air gap turns with the rotor. The main causes are: rotor outer diameter is not concentric, rotor thermal bent, bearing problems, rotor or load imbalance.

Mechanical problems such as rotor misalignment and imbalance can be also inferred in the low spectrum through the analysis of the rotational frequency sidebands. Figure 13 shows this pattern, where f_r is rotational frequency.

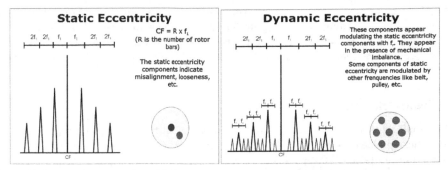

Figure 12. Static and Dynamic Eccentricities patterns

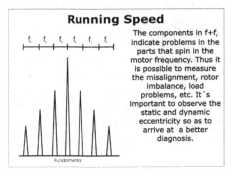

Figure 13. Rotational frequency pattern

5.2. Stator failure patterns

Most induction motor stator failures are related to the windings. The occurrence of failures in the stator core is less frequent. In spite of being rare, this last problem can cause considerable damages to the machine (Borges da Silva et al., 2009).

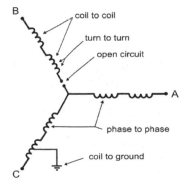

Figure 14. Stator winding failure modes

The failures related to the stator windings present a diversified set of possible manifestations according to the Figure 14. It is possible to notice their simultaneous occurrence. There are MCSA patterns for the detection of these failures, but EPVA is the most recommended technique to detect electrical imbalance in motors without direct torque control.

5.3. Bearing failure patterns

The monitoring of bearing damages is very important in predictive maintenance program since these problems account for 40% of the total amount of failures in an induction motor (Schoen et al., 1995). Many papers have recommended current signature analysis for the diagnosis of bearing faults, although it is important to register that this is an area that can be more explored and improved, tracking earlier fault detection.

There are several causes for bearing damages. Since this is not the objective of this work, the chapter presents just the characteristic components of failure in the outer and inner races, and rolling elements. The pattern is given by the Figure 15; where FBPFO is the rolling element characteristic frequency, FBPFI is the inner race characteristic frequency, FBSF is the outer race characteristic frequency, FFTF is the cage characteristic frequency, PD is the bearing pitch diameter, BD is the ball bearing diameter, β is the contact angle, n is the number of rolling elements, and Fr is the rotational speed.

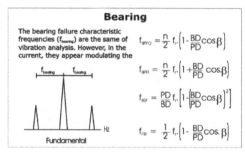

Figure 15. Bearing failure modes

5.4. Load failure patterns

The vast majority of the published papers about failure monitoring via current spectrum analysis presents the failure patterns related to broken bars and air gap eccentricity. This chapter presents a very meaningful contribution to the previous works since it adds new patterns related to the attached load. All the patterns have been tested, first through controlled laboratory tests and later through industrial cases. The failure patterns can be divided in three groups: motor failure, transmission system failure and attached load failure. By using the induction motor as a transducer, one can monitor the complete drive train, i.e., motor, transmission system and attached load, so as to increase the reliability of productive system.

5.4.1. Transmission System Failure

The MCSA monitors the frequency components related to pulleys (motor pulley and load pulley), belts and gear mesh. It has been observed that load problems can reflect in the transmission system frequency components. This characteristic is one more way of detecting mechanical load failures to be used in addition to the load characteristic frequency components.

1. **Pulleys**: by analyzing the rotational frequency one can detect problems related to the motor pulley. When there is no change in the speed, it is not possible to distinguish the damaged pulley from the healthy one since they have the same rotational frequency. But when a speed transformation is present, one can monitor the load pulley and the attached load through the pattern presented in Figure 16. In this case, f_{lf} is equal to f_{pulley}, and f_{pulley} is the load pulley characteristic frequency given by (45).

$$f_{pulley} = \frac{D_{motor\_pulley} \times f_r}{D_{load\_pulley}} \tag{45}$$

Where f_r is the rotational frequency, D_{motor_pulley} is the diameter of the motor pulley and D_{load_pulley} is the diameter of the load pulley. The sideband components of the fundamental are at $f_1 \pm f_p$.

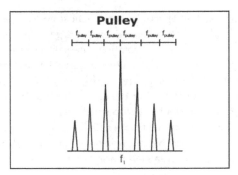

Figure 16. Load Pulley Pattern

The most common problems are eccentric pulley, pulley with mechanical looseness and unbalanced pulley. Problems related to the load can also reflect in the same frequencies. When this happens, the analyst himself must cross pieces of information from other spectrum regions so as to arrive at a reliable conclusion.

2. **Belts**: the first step when monitoring the belt characteristic frequency components is to calculate the belt frequency (f_b). In this case, f_{lf} is equal to f_b, and f_b is the belt characteristic frequency given by (46).

$$f_b = \frac{D_{motor\_pulley} \times \pi \times f_r}{L_{belt}} \tag{46}$$

Where L_{belt} is the belt length

This way the sideband components of the fundamental are at $f_1 \pm f_b$. After calculating this frequency, it is enough to follow the pattern presented in Figure 17 and follow up the tendency curve in order to diagnose problems in this transmission system element.

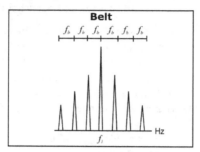

Figure 17. Belt Failure Pattern

Besides diagnosing problems such as loosen belt, broken belt or too taut belt, one can analyze problems originating in the load. In case of load failure, the vibration levels in the belts increase considerably and result in higher amplitudes for the belt characteristic frequencies.

3. **Gear Mesh**: in this case, two spectrum regions must be monitored. The first one, in a lower frequency band, shows punctual failure in the gear (for instance, a broken tooth). These frequencies are related to the rotational frequencies before and after the speed transformation. This way the sideband components of the fundamental are at $f_1 \pm f_{r1}$ and $f_1 \pm f_{r2}$ respectively. Where f_{r1} is the rotational frequency before the speed transformation and f_{r2} is the rotational frequency after the speed transformation. The second spectrum region of interest shows distributed failures in the gear. They are known as gear mesh frequency (f_g) and can be calculated by multiplying the rotational shaft speed by the gear teeth number Figure 18a illustrates this situation, and Figure 18b shows the sideband components of the fundamental are at $f_1 \pm f_g$.

$$f_g = n \cdot f_{r1} = N \cdot f_{r2} \tag{47}$$

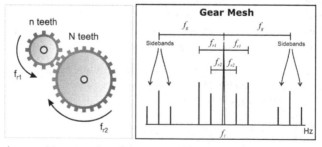

Figure 18. Gear features: (a) gear mesh, and (b) gear mesh failure pattern

5.4.2. Attached load failure

As seen previously, a load fault reflects in the motor stator current by means of torque oscillations. This chapter presents in this section three different kinds of loads and their respective patterns. Other load types result in different patterns but the fundamental sequence is always the same: define the characteristic frequencies from the constructive data, find their presence in the motor current signature due to torque oscillations from load faults, analyze the tendency curve and diagnose the fault.

1.   **Centrifugal Pumps:** for the analysis of centrifugal pumps one has to consider the pump rotational frequency (f_{r_pump}) and the vane pass frequency (f_{vp}) that is given by:

$$f_{vp} = n \cdot f_{r\_pump} \tag{48}$$

Where n is the number of pump vanes.

The analysis of the pump rotational frequency (f_{r_pump}) indicates problems related to misalignment or pump imbalance. In this case, f_{lf} is equal to f_{r_pump} and the sideband components of the fundamental are at $f_1 \pm f_{r\_pump}$. On the other hand, the increase of the amplitudes of vane passing frequency indicates problems inside the pump, such as vane deterioration. Now the sideband components of the fundamental are at $f_1 \pm f_{vp}$. Figure 19 shows the pattern for these frequencies.

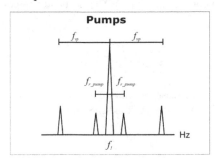

Figure 19. Centrifugal pump failure pattern

In addition to those frequencies one has to monitor the increase of saliencies close to the supply frequency. These frequencies are characteristic of pump signature and also can indicate pump problems.

2.   **Screw Compressor:** the complete set motor, gear mesh and screw compressor can be monitored by means of MCSA satisfactorily. The motor and the gear mesh can be analyzed according to the patterns presented previously. Figure 20a shows the scheme of a screw compressor. Where N is the motor gear teeth number, n is the compressor gear teeth number, L_m is the male screw lobules number, L_f is the female screw lobules number, F_r is the motor rotational frequency, F_{r1} is the male screw rotational frequency, F_{r2} is the female screw rotational frequency and F_p is the pulsation frequency. The screw compressor failure spectral pattern is presented in Figure 20b.

The screw compressor analysis takes into consideration three characteristic frequencies:

a.   Male screw rotational frequency: in this case, $f_{if} = f_{r1}$ and f_{r1} is the male screw rotational frequency given by (51). The sideband components of the fundamental are at $f_i \pm f_{r1}$.

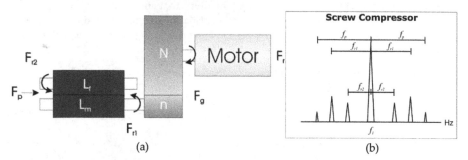

Figure 20. Screw compressor: (a) schematic and (b) failure spectral pattern

$$f_{r1} = \frac{N}{n} \cdot f_r \tag{49}$$

b.   Female screw rotational frequency: in this case, f_{if} is equal to f_{r2} and f_{r2} is the female screw rotational frequency given by (50). The sideband components of the fundamental are at $f_i \pm f_{r2}$.

$$f_{r2} = \frac{L_m}{L_f} \cdot f_{r1} \tag{50}$$

c.   Pulsation frequency: in this case, f_{if} is equal to f_p and f_p is the pulsation frequency given by (51). The sideband components of the fundamental are at $f_i \pm f_p$.

$$f_p = L_m \cdot f_{r1} = L_f \cdot f_{r2} \tag{51}$$

When the screw compressor has two stages, it is enough to apply the same reasoning for the second stage of compression. Since the speed transformations are different, the characteristic component of each stage can be separated in the spectrum.

3.   **Fans:** in the same way of pumps, fan failure analysis considers the fan rotational frequency and the blade passing frequency (f_{bp}):

$$f_{bp} = N_b \times f_{r\_fan} \tag{52}$$

Where N_b is the number of blades and f_{r_fan} is the fan rotational frequency.

Analyzing the rotational frequency (f_{r_fan}), problems related to misalignment or fan imbalance can be detected. When, f_{if} is equal to f_{r_fan} and the sideband components of the fundamental are at $f_i \pm f_{r\_fan}$. Also, the increase of the amplitudes of blade passing frequency indicates problems like blade deterioration or break. The sideband components of the fundamental are given by $f_1 \pm f_{bp}$. Figure 21 shows the fan failure patterns.

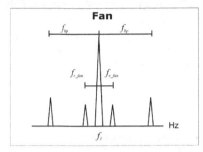

Figure 21. Fan failure pattern

6. Elements of a monitoring system for predictive maintenance

A sophisticated monitoring system can read the entrances of hundreds of sensors and execute mathematical operations and process a diagnosis. Currently, the diagnosis is gotten, most of the time, using artificial intelligence techniques (Lambert-Torres et al. 2009).

Considering the previous statements, a monitoring system can be divided in four main stages: (a) transduction of the interest signals; (b) acquisition of the data; (c) processing of the acquired data; and (d) diagnosis. Figure 22 presents a pictorial form of this process.

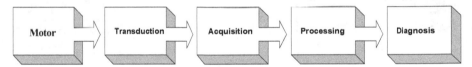

Figure 22. Steps of the Monitoring Process

6.1. Transduction

A transducer is a piece of equipment that has in its entrance an input value to be monitored (current, voltage, acceleration, temperature, etc), whereas in its output it has a signal that is conditioned and envoy to the acquisition system and processing. The main transducers used in the monitoring processing of electric machines are:

- For measurement of temperature: they are the three main methods of measurement of temperature: thermocouple, thermister, and RTD (Resistance Temperature Detection).
- For the measurement of vibration: two types of transducers for the vibration analysis exist: the absolute transducers or with contact and the relative ones or without contact. The absolute transducers measure the real movement of the machine, whereas the relative ones measure the movement of an element of the machine in relation to the other element. The accelerometer is the main and more used existing absolute sensor in the market.

- For measurement of force: the most common is the strain gauge, that it is a device that understands a resistance that has its modified size and transversal area in function of the application of a force. Then, the force can be measured through the variation of the resistance.
- For measurement of electric and magnetic values: the electric values are measured from transforming of voltage and current those always are presented as part of the protection system. However, it can still have the necessity an extra measure, the density of magnetic flow in the machine, using itself a hall-effect device.

6.2. Data acquisition

The data acquisition is a stage with fundamental importance; because it needs to guarantee the integrity and precision of the collected data. The precision of the data demanded of the acquisition is determined by the future mathematical manipulations that are applied to the data set. The collection and the transmission of the data must be made in order to minimize to the maximum the effect of the noise, being become the sufficiently consistent data. In complex systems with many entrances, it is oriented that the processing system is remote, that is, located to a certain distance of the inspected process. Figure 23 presents an example where some motors are being monitored. A group of adjacent machines is connected to a point of collection of data that digitalize the signal and sends for the remote central office of processing and diagnosis.

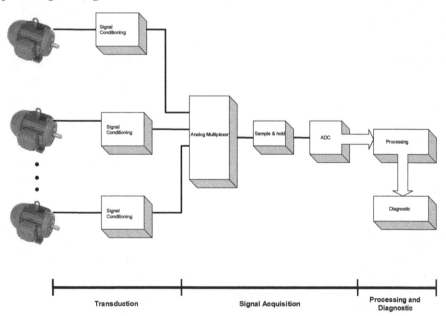

Figure 23. Example of a Monitoring System

The multiplexer is essential when a great number of channels must be monitored. Moreover, it also is recommended for a small number of channels, since it allows the use of only one converter A/D. Already converter A/D is the heart of the acquisition system and must be specified in function of the time of conversion and precision.

6.3. Processing

The task of the processing must be to catch the collected data and to manipulate them and/or to transform them, aiming at the agreement of these for the system of diagnosis in a faster and easier form. The processing can be done on-line or off-line. The choice depends on the process that are being monitored and on the speed with that the characteristics of interest of this process modify themselves.

There are different techniques of processing to monitor electric machines. One of the simplest of them, it examines the amplitude of the signal of entrance of the function in the time, and compares it with a predetermined value. Elaborated techniques are currently possible due to the new computers, such as: spectral analysis, correlation, averages, cepstral, envelope analysis, etc.

6.4. Diagnosis

Diagnosis is the part most critical of the system, because it involves decisions and consequently money. Currently, many techniques of artificial intelligence as expert systems and neural nets are being used (Lambert-Torres et al., 2009).

7. Implementation in a real-case predictive maintenance

A Brazilian petroleum company has decided to implement electrical signature analysis through a remote condition monitoring system named Preditor (PS Solutions, 2011). The communication is based on Ethernet network. Each hardware has been plugged in this network has an IP address and through the motor configuration the software knows exactly where each signal comes from. This way it is possible to monitor the motor condition from a remote office with a group of expert analysts or to count on the automatic support of the software.

Among the induction motors monitored, an example of electrical imbalance was chosen. Motor nameplate features are 250 CV, 2400 V, 70 A, 505 RPM, 14 poles, and attached to a reciprocating compressor. The remote system software has indicated electrical imbalance based on EPVA signature. Figure 24 presents the stator electrical imbalance signature and tendency curve for this motor.

One can observe from the figure above that the electrical imbalance was around 5.7%. For an idea of magnitude, all the other motors presented an electrical imbalance around 1%. The motor history was tracked and the maintenance department detected a set of defective coils in one phase. These coils were by-passed, which caused the imbalance, as shown in Figure 25.

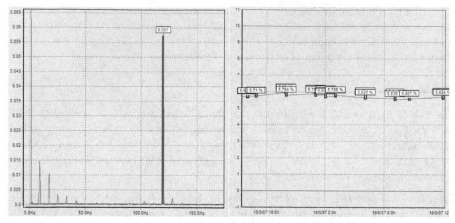

Figure 24. Stator electrical imbalance signature and tendency curve

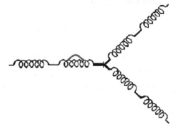

Figure 25. Set of defective coils by-passed

After all the implementing job, one can say that the remote system based on electrical signature analysis is an effective alternative for rotating machines monitoring since the system fits the refineries safety rules. It still allows the non-intrusive monitoring, avoiding exposing the workers to electrical shock and arcs, confined spaces and eliminating the necessity of job permissions and risk analysis for signal acquiring (which implies in cost reduction). The electrical failure dynamic monitoring presents a good potential to increase the industries process reliability. Besides, the techniques also allow the tracking of mechanical components, which is an interesting tool to detect mechanical faults in machines located in places of difficult access.

In 2006, a petroleum refinery experienced an unplanned outage in its Coker Unit caused by the breakage of some rotor bars in the induction motor of the decoking pump which damaged the rotor and the stator of the motor as can be seen in the picture below. The damaged motor had the following features: Poles – 4, Rated Power – 1700 kW, Rated Voltage – 13.8 kV, Shaft Height – 500 mm, and Hazardous Area – Free zone. Although it is not possible to operate without the decoking pump, there is not a standby motor because of its high reliability and cost. Figure 26 shows the stator and rotor damages.

Figure 26. Stator and Rotor Damages

After the event, the motor was sent to be repaired, but the first information was that it would take 70 days to be fixed. Since this deadline would compromise the refinery production plan, the refinery's maintenance team started looking for a similar motor. In normal conditions, it was not possible to find a better solution, than to wait for 70 days (considering the purchase of a new motor it would take, at least, 6 months). Luckily, a motor was found in a factory with the following features: Poles – 4, Rated Power – 1656 kW, and Rated Voltage – 4.16 kV, Shaft Height – 450 mm, and Hazardous Area – Free zone.

Considering that the refining process is based on pumps and compressors, the engineers noticed that the unique parameter that should be exactly the same was the number of poles. To the others, the following analysis was done:

- Rated Power – Since the original motor does not operate at its rated power, it was possible to use the similar motor;
- Rated Voltage – the refinery had a voltage transformer in stock (4.16/13.8 kW), that could be used to supply the rated voltage to the similar motor;
- Shaft Height – the original motor shaft was higher than the similar one, but this could be solved easily by adapting the skid.

Besides, considering that the decoking pump had been installed in a non hazardous area, the similar motor completely met the requirement to be installed. Then, after a short negotiation, an agreement was made between the oil company and the motor manufacturer, where the similar motor was rented to be adapted, while the manufacturer made another motor to replace the original one. While the similar motor was in its way to the refinery, all possible and necessary electrical and mechanical work to fit this motor to the site was in process. When the similar motor arrived, the maintenance team spent only one day to replace the motor. Six days after the outage, the Coker Unit started over.

8. Cost analysis

Based on the Brazilian Petroleum Company experience reported above, in terms of costs, it is very easy to demonstrate the benefits of having an ESA system installed together with a motor management.

Considering that 1 day without production means losses of US$ 300,000.00, we would have US$ 21,000,000.00 in 70 days. However, as we found a motor to be adapted, we had just 6 days of losses (US$ 1,800,000.00). If we had an ESA System installed monitoring this motor, we could realize in advance that the motor was developing a failure. As we said before, some refineries have similar motors that could be adapted. So, in that case, it would be possible to plan the replacement, sending the motor, and making the adaptations and stopping the production for only 1 day, i.e. losses of US$ 300,000.00.

9. Conclusions

The industries currently look for products and outside services for predictive maintenance. In many cases, the outside service company or even the industrial plant predictive group make mistakes that can compromise the whole condition monitoring and failure diagnosis process. In this increasing demand for prediction technology, a specific technique referred as Electrical Signature Analysis (ESA) is calling more and more attention of industries.

Considering this context, the presented chapter intends to disseminate important concepts to guide companies that have their own predictive group or want to hire consultants or specialized service to obtain good results through general predictive maintenance practices and, especially through electrical signature analysis.

The result of the proposed discussion in this chapter is a procedure of acquisition and analysis, which is presented at the end of the chapter and intends to be a reference to be used by industries that have a plan to have MCSA as a monitoring condition tool for electrical machines.

Author details

Erik Leandro Bonaldi, Levy Ely de Lacerda de Oliveira,
Jonas Guedes Borges da Silva and Germano Lambert-Torres
PS Solutions, Brazil

Luiz Eduardo Borges da Silva
Itajuba Federal University, Brazil

Acknowledgement

The academic authors would like to express their thanks for CNPq, CAPES, FINEP and FAPEMIG for support this work.

10. References

Benbouzid, M.H. (2000). A Review of Induction Motors Signature Analysis as a Medium for Faults Detection, *IEEE Transactions on Industrial Electronics*, Vol.47, No.5, (October 2000), pp. 984-993, ISSN 0278-0046.

Bonaldi, E.L., Borges da Silva, L.E., Lambert-Torres, G. de Oliviera, L.E.L. (2003). A Rough Sets Based Classifier for Induction Motors Fault Diagnosis, *WSEAS Transactions on Systems*, Vol.2, No.2, (April 2003), pp. 320-327, ISSN 109-2777.

Bonaldi, E.L., de Oliviera, L.E.L., Lambert-Torres, G. & Borges da Silva, L.E. (2007). Proposing a Procedure for the Application of Motor Current Signature Analysis on Predictive Maintenance of Induction Motors, *Proceedings of the 20th International Congress & Exhibition on Condition Monitoring and Diagnosis Monitoring Management - COMADEM 2007*, Faro, Portugal, Jun. 13-15, 2007.

Bonnett, A.H., & Soukup, G.C. (1992). Cause and Analysis of Stator and Rotor Failures in Three-Phase Squirrel-Cage Induction Motors, *IEEE Transactions on Industrial Electronics*, Vol.28, No.4, (July/August 1992), pp. 921-937, ISSN 0278-0046.

Borges da Silva, L.E., Lambert-Torres, G., Santos, D.E., Bonaldi, E.L., de Oliveira, L.E.L. & Borges da Silva, J.G. (2009). An Application of MSCA on Predictive Maintenance of TermoPE's Induction Motors, *Revista Ciências Exatas*, Vol. 15, No. 2, (July 2009), pp. 100-108, ISSN 1516-2893.

Cardoso, A.J.M. & Saraiva, E.S. (1993). Computer-Aided Detection of Airgap Eccentricity in Operating Three-Phase Induction Motors by Park's Vectors Approach, *IEEE Transactions on Industry Applications*, Vol.29, No.5, (Sept/Oct 1993), ISSN 0093-9994.

Lambert-Torres, G., Bonaldi, E.L., Borges da Silva, L.E. & de Oliviera, L.E.L. (2003). An Intelligent Classifier for Induction Motors Fault Diagnosis, *Proceedings of the International Conference on Intelligent System Applications to Power Systems - ISAP'2003*, Paper 084, Lemnos, Greece, Aug. 31 – Sept. 3, 2003.

Lambert-Torres, G., Abe, J.M., da Silva Filho, J.I. & Martins, H.G. (2009). *Advances in Technological Applications of Logical and Intelligent Systems*, IOS Press, ISBN 978-1-58603-963-3, Amsterdam, The Netherlands.

Legowski, S.F., Sadrul Ula, A.H.M., & Trzynadlowski, A.M. (1996). Instantaneous Power as a Medium for the Signature Analysis of Induction Motors. *IEEE Transactions on Industry Applications*, Vol.32, No.4, (July/August 1996), pp. 904-909, ISSN 0093-9994.

PS Solutions. (October 2011). Predictor, Available from www.pssolucoes.com.br, visited on 22/10/2011.

Schoen, R.R., Habetler, T.G., Kamram, F. & Bartheld, R.G. (1995). Motor Bearing Damage Detection Using Stator Current Monitoring, *IEEE Transactions on Industrial Electronics*, Vol.31, No.6, (Nov/Dec 1995), pp. 1274-1279, ISSN 0278-0046.

Tavner, P.J., Ran, L., Penman, J. & Sedding, H. (1987). *Condition Monitoring of Rotating Electrical Machines*, The Institution of Engineering and Technology – IET, 2nd Edition, ISBN 978-0863417412, London, UK.

Thomson, W.T., & Fenger, M. (2001). Current Signature Analysis to Detect Induction Motor Faults, *IEEE Industry Applications Magazine*, Vol.7, No.4, (July 2001), pp. 26-34, ISSN 1077-2618.

Thorsen, O.V. & Dalva, M. (1999). Failure Identification and Analysis for High-Voltage Induction Motors in the Petrochemical Industry, *IEEE Transactions on Industry Applications*, Vol.35, No.4, (July/August 1999), pp. 810-817, ISSN 0093-9994.

Permissions

The contributors of this book come from diverse backgrounds, making this book a truly international effort. This book will bring forth new frontiers with its revolutionizing research information and detailed analysis of the nascent developments around the world.

We would like to thank Rui Esteves Araújo, for lending his expertise to make the book truly unique. He has played a crucial role in the development of this book. Without his invaluable contribution this book wouldn't have been possible. He has made vital efforts to compile up to date information on the varied aspects of this subject to make this book a valuable addition to the collection of many professionals and students.

This book was conceptualized with the vision of imparting up-to-date information and advanced data in this field. To ensure the same, a matchless editorial board was set up. Every individual on the board went through rigorous rounds of assessment to prove their worth. After which they invested a large part of their time researching and compiling the most relevant data for our readers. Conferences and sessions were held from time to time between the editorial board and the contributing authors to present the data in the most comprehensible form. The editorial team has worked tirelessly to provide valuable and valid information to help people across the globe.

Every chapter published in this book has been scrutinized by our experts. Their significance has been extensively debated. The topics covered herein carry significant findings which will fuel the growth of the discipline. They may even be implemented as practical applications or may be referred to as a beginning point for another development. Chapters in this book were first published by InTech; hereby published with permission under the Creative Commons Attribution License or equivalent.

The editorial board has been involved in producing this book since its inception. They have spent rigorous hours researching and exploring the diverse topics which have resulted in the successful publishing of this book. They have passed on their knowledge of decades through this book. To expedite this challenging task, the publisher supported the team at every step. A small team of assistant editors was also appointed to further simplify the editing procedure and attain best results for the readers.

Our editorial team has been hand-picked from every corner of the world. Their multi-ethnicity adds dynamic inputs to the discussions which result in innovative outcomes. These outcomes are then further discussed with the researchers and contributors who give their valuable feedback and opinion regarding the same. The feedback is then collaborated with the researches and they are edited in a comprehensive manner to aid the understanding of the subject.

Apart from the editorial board, the designing team has also invested a significant amount of their time in understanding the subject and creating the most relevant covers. They scrutinized every image to scout for the most suitable representation of the subject and create an appropriate cover for the book.

The publishing team has been involved in this book since its early stages. They were actively engaged in every process, be it collecting the data, connecting with the contributors or procuring relevant information. The team has been an ardent support to the editorial, designing and production team. Their endless efforts to recruit the best for this project, has resulted in the accomplishment of this book. They are a veteran in the field of academics and their pool of knowledge is as vast as their experience in printing. Their expertise and guidance has proved useful at every step. Their uncompromising quality standards have made this book an exceptional effort. Their encouragement from time to time has been an inspiration for everyone.

The publisher and the editorial board hope that this book will prove to be a valuable piece of knowledge for researchers, students, practitioners and scholars across the globe.

List of Contributors

Alfeu J. Sguarezi Filho
Universidade Federal do ABC, Brazil

José Luis Azcue and Ernesto Ruppert
School of Electrical and Computer Engineering, University of Campinas, Brazil

Manuel A. Duarte-Mermoud & Juan C. Travieso-Torres
Department of Electrical Engineering, University of Chile, Santiago, Chile

Sebastien Solvar, Malek Ghanes, Leonardo Amet, Jean-Pierre Barbot and Gaëtan Santomenna
ECS - Lab, ENSEA and GS Maintenance, France

José Luis Azcue and Ernesto Ruppert
University of Campinas (UNICAMP), Brazil

Alfeu J. Sguarezi Filho
CECS/UFABC, Santo André - SP, Brazil

Pedro Melo
Polytechnic Institute of Porto, Portugal

Ricardo de Castro and Rui Esteves Araújo
Faculty of Engineering – University of Porto, Portugal

Raúl Igmar Gregor Recalde
Engineering Faculty of the National University of Asunción, Department of Power and Control Systems, Asunción-Paraguay

Marcin Morawiec
Gdansk University of Technology, Faculty of Electrical and Control Engineering, Poland

Ouahid Bouchhida
Université Docteur Yahia Farès de Médéa, Département Génie Electrique, Algérie

Mohamed Seghir Boucherit
Ecole Nationale Polytechnique, Département Génie Electrique, Algérie

Abederrezzek Cherifi
IUT Mantes-en-Yvelines, France

Ivan Jaksch
Technical university of Liberec, Czech Republic

Erik Leandro Bonaldi, Levy Ely de Lacerda de Oliveira, Jonas Guedes Borges da Silva and Germano Lambert-Torres
PS Solutions, Brazil

Luiz Eduardo Borges da Silva
Itajuba Federal University, Brazil